Glacial and Periglacial Geomorphology

Second edition
Volume 2

Periglacial Geomorphology

Clifford Embleton

Reader in Geography, University of London King's College

Cuchlaine A. M. King

Professor of Physical Geography, University of Nottingham

A HALSTED PRESS BOOK

John Wiley & Sons
New York

First published 1975 by
Edward Arnold (Publishers) Ltd
London

This book is a fully revised edition of Part IV of
Glacial and Periglacial Geomorphology first published 1968 by
Edward Arnold (Publishers) Ltd
Reprinted 1969, 1971, 1975

Published in the USA
by Halsted Press, a Division
of John Wiley & Sons, Inc
New York

Library of Congress Cataloging in Publication Data

Embleton, Clifford
  Glacial geomorphology

  (Their Glacial and periglacial geomorphology; v. 1)
  'A Halsted Press book'
  Includes bibliographies and index
  1. Glacial landforms.   I. King, Cuchlaine, A. M. joint author.   II. Title
GB581.E452   1975   vol. 1   551.3'1'08s
ISBN 0–470–23892–5   [551.3'1]   75–14188
ISBN 0–470–23893–3 pbk

Printed in Great Britain

Preface

The first edition of this book was published (1968) in a volume which combined a treatment of both glacial and periglacial geomorphology. Two reasons prompted us to separate these topics when the book fell due for revision. First, because of the enormous expansion of research in these fields, we felt that an adequate account at this level demanded a lengthier treatment; this in turn meant a division into two volumes if the work was not to become unduly bulky. Secondly, although there are close links between these two branches of geomorphology, they are not so close that separation involves any great disadvantage, nor does it involve any significant repetition. The majority of research workers in these fields have concentrated their attention on one or the other, rarely both, and some schools of geomorphology have developed particularly strong interests in one, as for instance, the long tradition of Polish research in periglacial geomorphology.

References are listed at the end of each chapter. The majority are to the most recently published literature. The authors are aware that inadequate reference is given to the voluminous Soviet and East European literature; one of the great problems facing geomorphologists (and other scientific workers) at the present time is that of language, though, increasingly, translations of some outstanding items are becoming available. In the reference lists, abbreviations of titles of periodicals are given in accordance with the *World List of Scientific Periodicals* (4th edition). In the case of books cited, place of publication is given only if the work is not published in London or New York.

All measurements used in the book are given in the metric system, using SI units.

The authors are grateful to many who have helped in the preparation of this book. Most of the illustrations have been prepared in the drawing office of the Department of Geography, King's College, London. We would like particularly to thank Mrs Anne Rogers for meticulous typing and help in checking proofs.

June 1974

CLIFFORD EMBLETON

CUCHLAINE A. M. KING

Contents

Plates

The 'subglacial' climate must be looked on as an optimum of destructive action.

(J. G. ANDERSSON, 1906)

1

The periglacial environment: an introductory survey

Freezing is the most important factor in a periglacial climate. (W. LOZINSKI, 1912)

The idea that distinctive phenomena were associated with a cold climatic environment induced by the presence of nearby glacial ice in Pleistocene times was formulated in the nineteenth century. Crystallization of thought on the matter was encouraged by J. Geikie's volume *Great Ice Age* (1874) in which were summarized the known occurrences of 'rubble drift' or 'head' in Britain. Geikie and others favoured an origin for these deposits under former cold conditions. In 1900, F. E. Matthes drew attention to a process, nivation, demanding a freeze-thaw snow climate but not actually involving glaciation. Six years later, J. G. Andersson reported on his visit to Bear Island in the Arctic, describing a hitherto unrecognized form of mass wasting which he termed 'solifluction', a process active under 'sub-glacial' climatic conditions. The adjective 'sub-glacial' was clearly ambiguous, and in 1909, W. Lozinski proposed the term 'periglacial'. Lozinski had studied mechanical weathering of sandstones in the Carpathians which seemed to require a former frost climate. His ideas, presented to the Eleventh International Geological Congress meeting at Stockholm in 1910 (Lozinski, 1912), stimulated discussion and awoke considerable interest in the subject of periglacial geomorphology, which was further intensified among the participants of the subsequent excursion to Spitsbergen. Unfortunately, Lozinski failed to define the term 'periglacial' explicitly; he, moreover, was mostly concerned with frost weathering and gave little consideration to the distinctive processes of nivation and solifluction recognized by Matthes and Andersson. Lozinski used the term periglacial to refer to areas lying near the margins of Pleistocene ice, and to the supposedly distinctive climatic conditions of these areas. By extension, the term came to be applied also to the phenomena and processes induced by such climatic conditions. To the range of periglacial phenomena so far recognized were added, in the next few years, equiplanation (D. D. Cairnes, 1912), altiplanation (H. M. Eakin, 1916), and frost-heaving (B. Högbom, 1914). Högbom's monograph on frost phenomena, with particular reference to Scandinavia and Spitsbergen, marked a milestone in the progress of periglacial studies, and is still a standard reference.

The term periglacial is widely used today, though its usage and varied application are far from satisfactory. The most acceptable application of the term is to a zone, albeit of indefinite width, peripheral to the glacial ice of today or of any phases of the Pleistocene. To speak of a 'periglacial climate', unless it is arbitrarily defined, is unsatisfactory, for conditions in the periglacial zone of present and Pleistocene ice sheets vary enormously from place to place. Today, for instance, there are very significant differences between north Greenland and Spitsbergen, though both lie in the same latitudes. W. E. Davies (1961) describes the former as a High Arctic desert, where precipitation is only about 50 mm a year; mean annual temperature at Thule is $-12°C$, and wind velocities may reach 40 m/s. Because of the dryness, many of the surface features commonly associated with permafrost in areas of moist or saturated soils are generally lacking. The formation of patterned ground is retarded or even arrested owing to the lack of moisture. At Hornsund in West Spitsbergen (Z. Czeppe, 1960), temperatures are high enough in summer for the ground to thaw out completely for 2 months of the year, while in spring and autumn there are numerous freeze-thaw cycles. Moisture is relatively abundant because of the more maritime climate and the thawing of ground ice.

Climatic contrasts in the periglacial zone were equally striking in Pleistocene glacial times. A severely cold dry climate characterized the Weichselian maximum in southern European Russia, for example, when great spreads of wind-blown loess accumulated beyond the ice margin, and must have been very different from the more maritime and probably 'Icelandic' type of climate found then in ice-free areas of southern Britain. Although large areas of the Pleistocene periglacial zone were certainly characterized by continuous or semi-continuous permafrost, there were also undoubtedly other parts whose climate was not too different from that of the present, as in the Wisconsinan ice-marginal areas of Illinois, Indiana and Ohio where tree growth may have continued close to the actual ice front (see p. 185). H. Poser (1948) distinguished six major regions of periglacial climate in Europe during the Würm, ranging from tundra (with maritime, intermediate and continental varieties) to 'forest climates' with or without permafrost. Moreover, the periglacial zone of the last glaciation in middle latitudes is not simply to be equated in terms of its range of climate with present-day high-latitude arctic regions, for as J. Büdel (1959) has pointed out, the former was characterized by higher angle sun and greater insolation, thus promoting more rapid snow melt, a greater depth of seasonal ground thaw, and possibly a greater drying-out of the ground in the late summer. Büdel attempts to distinguish such a régime as 'temperate periglacial'. The concept of differing periglacial climates is to be encouraged, as opposed to the often-expressed notion that there could be one periglacial climate experiencing low temperatures, many freeze-thaw fluctuations, and occasional strong winds blowing from a supposed glacial anticyclone.

L. C. Peltier (1950) attempted to define a periglacial morphogenetic region, which he took to be an area whose approximate range of mean annual temperature was $-15°$ to $-1°C$, whose mean annual precipitation was from 120 to 1400 mm, where strong mass movements and occasionally strong winds prevailed, and where the action of running water was relatively minor. A modified version of this scheme was proposed by L. Wilson (1969), in which the temperature limits were approximately $-12°$ to $+3°C$

and the precipitation range 50–1250 mm. Wilson recognized a 'periglacial climate-process system' in which solifluction, running water and frost action were said to be dominant, its extent being equated with Köppen's ET (tundra), EM and Dc (humid microthermal) zones. There are many geomorphologists who believe, however, that simple correlations between climate and geomorphological activity are unjustified (see D. R. Stoddart, 1969, for a useful review). Such generalized attempts at regionalization conceal great diversity of form and process, and it is seriously open to question whether such climatic parameters as mean annual temperature or mean annual precipitation are particularly meaningful for geomorphological analysis. Moreover, the positions and climates of the periglacial zone through the Pleistocene up to the present have changed radically. The periglacial zone is not a static phenomenon, nor are its characteristics everywhere the same.

Many writers, from F. E. Zeuner (1945) to J. Tricart (1967, 1969) have argued that permafrost is the primary characteristic of the periglacial zone. It is certainly a convenient definition to equate periglacial areas with extent of permafrost (B. A. Kennedy, 1969), but it is impossible to be so rigid in practice, as Tricart recognizes, for on the one hand, some minor periglacial features lie outside its limits yet, on the other hand, some areas lacking permafrost (e.g. seasonally unfrozen areas) nevertheless deserve to be included within it. L.-E. Hamelin (1961) extended the term periglacial environment to cover all areas in which frost action is or has been an important factor; this is a much wider usage of the term periglacial than most would probably accept. D. L. Linton (1969) suggests that the solution may be to abandon the term periglacial altogether, because of the imprecision with which it has been used, and instead to use more specific descriptions, such as permafrost or ground-ice environment. But the term periglacial is so strongly entrenched in the past and current literature of so many countries and languages that it seems improbable that Linton's suggestion will be adopted. The term remains a convenient one under which to group certain features and processes related to a range of climates characteristic of a broad periglacial zone, either now or in the past, and will be so used in this book.

There have been several attempts to recognize a vertical zonation of periglacial activity and phenomena, analogous to lateral variations within the periglacial zone. Reviewing data from the Alps, the Pyrenees and the Apennines, P. W. Höllermann (1967) tabulated altitudinal data for the snow-line and the timber-line, and the lower limits of gelifluction and ground sorting. Gelifluction limits were not difficult to identify, though variable, but defining the lower limit of sorting presents more problems. H. Stingl (1969) described a lower zone of gelifluction lobes in the eastern Alps and a higher zone of sorted forms. In south Swedish Lapland, S. Rudberg (1972) differentiates three altitudinal zones. Above the timber-line is a region of gelifluction lobes with turf-banked fronts and gliding boulders. Above this is a zone characterized by stone-banked terraces and stone stripes; higher still is the 'frost-shatter zone' with much bare rock, extensive blockfields and patterned ground.

Terminology

'Periglacial terminology is irrational, imprecise, incomplete and non-systematic'

(Hamelin and F. A. Cook, 1967). Many different terms have been used for a single phenomenon or process; some have been introduced with a particular theory in mind; and sometimes the same term has been used by different workers to describe dissimilar forms. The problem is compounded by language differences, and a useful review of some European language equivalents for periglacial terms is given by A. Dylikowa and J. Olchowik-Kolasinska (1956). Hamelin and Cook's *Illustrated glossary of periglacial phenomena* (1967) is particularly valuable for its bilingual (French–English) approach.

An early attempt to rationalize periglacial terminology was that of K. Bryan (1946), who proposed an entirely new set of terms, grouped under the heading of 'cryopedology', the name proposed for a sub-discipline dealing with frozen ground phenomena. Some of these terms will be introduced in subsequent chapters. Many of Bryan's terms have never been widely accepted, for there are often simpler and unambiguous alternatives: 'congelifraction' is a cumbersome replacement for 'frost action'; 'permafrost', though etymologically unsound, is too well established to be supplanted by 'pergelisol'; and the term 'active layer' is much more intelligible than 'mollisol'. While it is always desirable that terminology should be logical and scientifically based, there is an equally great need to avoid unnecessary proliferation of jargon.

Some of Bryan's recommendations, and derivatives from them, are, however, useful. The prefix 'cryo' (Gr. *kruos*, frost) effectively qualifies certain cold-climate forms and processes. 'Cryonival' indicates conditions in which freeze-thaw is dominant and snowmelt provides the water. 'Cryoturbation' (preferable to 'congeliturbation') covers processes that have no more satisfactory name in English than 'frost-churning' and that give rise to such important sub-surface features as involutions and injection plugs. 'Cryoplanation' comprises processes leading to overall reduction of relief under cold conditions.

1 Periglacial weathering processes

Freeze-thaw action is undoubtedly the most important process of rock weathering in the periglacial zone; it is the primary agent responsible for such features as talus accumulations and blockfields, and the breakdown of debris into particles fine enough to be handled by running water and the wind. The process, however, is by no means a simple one, nor is it fully understood. O. R. Grawe (1936) was one of the first to cast doubt on the simple idea that expansion of water to form ice would inevitably result in rock shattering. Many complex pressure-temperature relationships are involved in the water-ice change. If water is present in a completely closed cavity within rock, and if there are no impurities (including air) in the water, then a fall of temperature of the water to $-22°C$ accompanied by its conversion to crystalline ice will produce a theoretical maximum force of about $2100 \, kg/cm^2$ (P. W. Bridgman, 1912), though in fact practically no rocks could withstand a tensile stress of even one-tenth of this amount. In practice, one is never dealing with a completely closed cavity, otherwise there would be no means of entry for the water in the first place. It is possible, however, that in a wedge-shaped water-filled crack, rapid cooling from outside would cap the crack with ice and convert it into a closed system, though the expansive force of ice crystal growth might be relieved by deformation of the ice rather than further rupturing of the rock. The temperatures required to produce freezing of moisture in rocks are often much lower than expected, for the water contains air and impurities, latent heat must be removed, and water may be supercooled in some circumstances. Other factors that must be considered in connection with freeze-thaw rock weathering are the intensity of freezing, its duration, and the rate of freezing. D. C. Connell and J. M. C. Tombs (1971) attempted to measure experimentally the pressures exerted by ice-crystal growth. The maximum pressure recorded was only 0·2 bar (c. 200 g/cm²) but this, they considered, was unlikely to be the maximum attainable and they maintained that the process was probably an important one in frost shattering. The thermodynamics of

Plate I
Frost-shattered rock fragment on an Arctic beach. Note the angular/sub-angular nature of the beach material. (C.E.)

frost damage to porous solids (D. H. Everett, 1961) are complex and incompletely understood. A. Falconer (1969) suggests that the clay mineralogy is important in the break-up of rock. In experiments, he demonstrated three categories of susceptibility to frost damage: sound rocks, in which 50–100 per cent of the contained water froze at $-7°$ to $-10°C$, and in which no further freezing took place down to $-20°C$; unsound rocks, in which less than 50 per cent of the contained water froze with temperatures as low as $-40°C$; and there were eleven samples in which no freezing took place. Unsoundness is apparently not dependent on the freezing pressure. The unsound rocks contained clay on to which water molecules were absorbed; where spaces are small, the water molecules become so arranged that they do not freeze, but appear to be instrumental in rock disintegration as repulsive forces develop.

Many workers have investigated the relationships between the effectiveness of freeze-thaw as an agent of weathering in a given region, and the climatic régime of that region. Unfortunately, climatic data are often inadequate: for instance, most of the temperature records normally available are of air temperatures measured in a standard screen at intervals which may be too infrequent, whereas much more relevant to freeze-thaw action would be detailed records of ground temperature fluctuations. Air temperatures provide no more than a rough guide to ground temperatures, for ground temperatures are affected by insolation, the presence of water, ice, or snow, and rates of heat absorption and radiation. Hillslopes in sunshine will experience melting long before the air temperature rises above freezing; alternate sunshine and cloud may even themselves induce freeze-thaw in critical conditions, while dark-coloured rocks may retain heat to melt ice even when the air temperature has fallen below freezing. G. Taylor (1922) observed melting snow on boulders in Antarctica when the air temperature was only $-16 \cdot 5°C$. A. L. Washburn (1969) in the Mesters Vig district of north-east Greenland

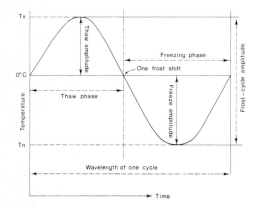

Fig. 1.1
Terminology associated with frost (freeze-thaw) cycles (K. Hewitt, *Can. Geogr.* 1968)

measured, at his experimental site 16, a temperature of $38 \cdot 4°C$ on a grus surface when the maximum air temperature recorded at a meteorological station $2 \cdot 5$ km away was $8°C$.

Fig. 1.1 defines a single complete freeze-thaw cycle. Washburn (1969), recognizing that depression of the temperature simply to $0°C$ (or to any other lower temperature down to a few degrees below zero) is no necessary guarantee that freezing has actually occurred, prefers the term 'zero-degree cycle', which *may* not be a freeze-thaw cycle but which at least can be simply and accurately defined from a set of temperature measurements.

Many workers have used air temperature records as a first approximation in studying the freeze-thaw process, and taken temperatures a few degrees above and below $0°C$ as the points of effective melt and freeze respectively. An example is J. K. Fraser's work (1959) on freeze-thaw cycles in Canada, which confirmed the suggestions made long ago by Högbom (1914) and E. K. Leffingwell (1919, p. 176) that high-latitude climates are usually unfavourable for freeze-thaw weathering. F. A. Cook and V. G. Raiche (1962) made an intensive study of freeze-thaw cycles at Resolute, N.W.T. (latitude $74°N$.), which, lying some 250 km from Devon Island ice cap, may fairly be said to lie in the present-day periglacial zone. Table 1.1 enumerates some of their findings. The small number of freeze-thaw cycles is striking (nearly all occur in May and June), as is the large number of mostly small amplitude crossings of the freezing line which may be too small to be significant in the weathering process. As expected, the ground surface experiences the most freeze-thaw cycles. Cook and Raiche contend that freeze-thaw cycles in this and probably many other parts of the Arctic are far less important than was once thought, a conclusion which has been recently supported by other investigators such as J. T. Andrews (1961, 1963a). Fraser (1959) found it difficult to

Table 1.1 Number of freeze-thaw cycles at Resolute, 1 May–30 September 1960

Level of temperature record	Range −2 to +1°C	Range −4 to +2°C	Frost-change days (at least one crossing of the freezing line)	Number of crossings of the freezing line
Stevenson screen	9	3	27	194
Stevenson screen, average for 1948–59	5·5	1·4	37·8	
Ground	18	7	44	170
−2·5 cm	1	0	13	16
−10 cm	0	0	8	24
−20 cm	0	0	9	8

avoid the conclusion that the abundance of apparently frost-riven rock in northern Canada is related not to the present climate but to conditions in the Pleistocene more favourable here for freeze-thaw action.

The data just presented for Resolute are probably typical of a present-day continental arctic climate. For comparison, Fig. 1.2, compiled from observations by Z. Czeppe (1960), illustrates a maritime arctic climate at Hornsund in Vest-Spitsbergen. The number of freeze-thaw cycles is greater than in the case of Resolute, and at ground

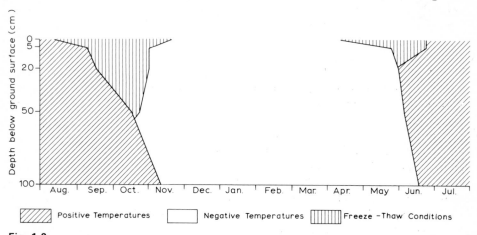

Fig. 1.2
Annual temperature régime recorded at Hornsund, Spitsbergen, for the ground surface and depths up to 100 cm (Z. Czeppe, *Bull. Acad. pol. Sci.*, 1960)

level, the total length of the two seasons when temperatures are frequently changing across the freezing line is over 5½ months. This type of temperature régime is clearly much more favourable for mechanical frost weathering, but even so, it should not be thought that such conditions represent the optimum. C. Troll's world survey (1944), though based on air temperature records, showed that the greatest frequency of freeze-thaw cycles is associated with low-latitude high-altitude areas, as in the Andes of Peru, and not, as Peltier (1950) thought, in the humid subarctic regions. However, air temperatures frequently crossing the freezing line are not the sole requirement for effective freeze-thaw action; moisture must also be available, and must be able to enter crevices or pore-spaces of suitable shape and size in the rocks. Availability of moisture is not adequately reflected in precipitation data from rain or snow gauges, as snow melt, fog and dew may be just as important as measurable precipitation in wetting the rocks.

Data on freeze-thaw cycles in high tropical and sub-tropical mountains are rarely available. K. Hewitt's (1968) records in the Karakoram Himalaya are therefore of

Table 1.2 Freeze-thaw cycles in the Karakoram Himalaya (K. Hewitt, 1968)

Month	Number of daily freeze-thaw cycles	Maximum amplitude of temperature change, °C
November	25	13·3
December	—	—
January	—	—
February	18	13·6
March	27	14·4
April	5	9·6

great interest (Table 1.2). Taken at a height of 3000 m, latitude 35°N, they show the importance of frequent and quite large diurnal freeze-thaw cycles in the sub-nival zone of the Karakoram. Particularly important from the point of view of frost action are the cycles in late winter and spring when more moisture is available.

There is still doubt as to whether rapid freezing of moisture in rocks is more effective than slow freezing, or whether intense cold during the period of freezing is more potent than temperatures just a few degrees below freezing. Numerous laboratory experiments on freeze-thaw action have been undertaken, but because of the variety of factors involved (rock type, amount of moisture, cycle amplitude, number of cycles, etc.) the experiments are never strictly comparable and it is not easy, therefore, to make valid generalizations. D. W. Kessler et al. (1940) used samples of water-soaked granite and relatively short 8-hour freeze-thaw cycles ranging between +20°C and −12°C. Five thousand cycles produced no significant rock damage. S. Wiman (1963) subjected various rock samples partially immersed in water to freeze-thaw cycles of two types: (a) 'Icelandic' type, with a daily periodicity (the temperature graph crossing freezing-point twice every 24 hours), a minimum of −7°C, and a maximum of +6°C; (b) 'Siberian' type, with a four-day cycle, a minimum of −30°C, and a maximum of +15°C. Both cycles were run for 36 days (i.e. 36 Icelandic cycles, and 9 Siberian cycles). Only slight disintegration of the samples was obtained, but the results suggested that the Icelandic type was more effective, unlike the conclusions of J. Tricart (1956), who found that over the same period, 8 Siberian-type cycles were more effective than 25 Icelandic-type cycles.

The results of experiments by A. S. Potts (1970) generally supported Wiman's conclusions. He used the same types of freeze-thaw cycle, except that the amplitude of the Icelandic type was slightly extended to range from $-8°C$ to $+8°C$. The cycles were run for longer periods—200 days in the case of the Icelandic type, and 400 days ($=100$ cycles) for the Siberian type. Two specimens of each rock type were used, one half-immersed in water, the other saturated but removed from the water; the former usually showed greater rates of disintegration. The products of shattering were removed and measured. As well as confirming Wiman's view that Icelandic-type cycles were more effective than the Siberian type, suggesting that intensity of freezing is less important than the number of freeze-thaw cycles, Potts also found that rates of shattering

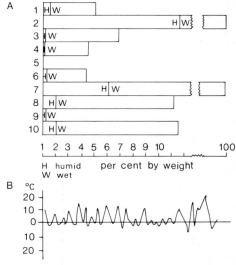

Fig. 1.3
Frost weathering of specimens of Tertiary basalt from northern Ireland. H=humid conditions, W=wet conditions (for definition see text). The bar graphs, **A**, show weight of disintegration products as a percentage of the weight of each specimen; the lower graph **B** shows the temperature fluctuations used in the experiment (G. R. Douglas, 1972, by permission of the author)

varied closely with rock type, igneous rocks proving most resistant and shales comparatively unresistant. Similarly, P. Rognon *et al.* (1967) found that granite and gneiss from the Vosges were relatively unaffected by 150 freeze-thaw cycles, while sandstone succumbed quite readily, some samples disintegrating completely.

G. R. Douglas (1972) has examined the response of basalt from northern Ireland to freeze-thaw action. Specimens were subjected to two sets of conditions:

1 'wet', in which air-dry samples were immersed in water up to one-third of their height,
2 'humid', where air-dry samples were covered in cotton-wool soaked in water.

Samples were subjected to 20 freeze-thaw cycles (approximately Icelandic type: see Fig. 1.3) and disintegration products weighed. Greatest yields were obtained from 'wet' conditions (W, Fig. 1.3), which were probably a closer match to actual field conditions where moisture is acquired by contact with ground water.

On the question of whether frost action can break down particles to clay or silt sizes, there is no unanimity of opinion. Potts (1970) obtained very little material finer than 60 μm and concluded that freeze-thaw action was probably not significant in producing the clay fraction of gelifluction deposits. On the other hand, Y. Guillien and J.-P. Lautridou (1970) found that freeze-thaw weathering of some Charente limestones did produce particles down to clay size. A problem at this micro-scale of particle disintegration is that it is virtually impossible to separate the possible effects of hydration from true freeze-thaw action.

Less work has been done on other processes of rock weathering in the periglacial zone. S. Taber (1943) and others acknowledge that sudden large falls of temperature will cause contraction cracking (sometimes audible) of ice-rich sediments, but low temperatures in themselves are not significant except in that they slow down rates of chemical action and also result in the elimination of free water to take part in chemical reactions. Chemical weathering under periglacial conditions has received relatively little attention until recently, though several writers such as C. Troll (1944) have remarked that ever since Lozinski linked the term periglacial with intensive frost weathering, there has been too much concentration on the latter at the expense of other possible processes. The notion that chemical weathering in arctic regions is negligible is now being gradually dispelled. Czeppe (1964) has examined sandstone outcrops in Spitsbergen which appear to be exfoliating in concentric shells to leave rounded core stones. His mineral analyses of the rocks and their shells prove successive chemical changes, and the splitting-off of the shells may be partly caused by expansion resulting from hydration or oxidation, though frost has undoubtedly aided the process once started. A. Cailleux (1962, 1968) described efflorescence and coatings of mirabilite (hydrated sodium sulphate), calcareous, ferruginous and silicic compounds, on rocks in the McMurdo Sound region of the Antarctic, and also the presence of rounded cavities (alveoles and taffoni) up to 0·1 m deep on the sides of boulders and as much as 2 m deep on some cliff faces. Although wind action must be allowed for in this arid area, the forms and positions of the cavities are incompatible with a hypothesis of wind abrasion; Cailleux maintains that wind action does not create them but certainly helps to keep them clean of weathered debris. He suggests that both freeze-thaw action and alternate solution and recrystallization of salts, principally mirabilite, are responsible for the etching of the rock surfaces, but concluded that, overall, chemical weathering was still subordinate to mechanical disintegration in this area.

Further studies in the McMurdo Sound area have been made by M. J. Selby (1972). He elaborates on the detailed sculpture of joint blocks by taffoni, honeycomb forms and curved shell-like surfaces. A ferric oxide varnish on the most exposed surfaces of joint blocks is certainly a product of chemical weathering and, once formed, helps to protect these surfaces while the inner core continues to rot away. Salt weathering is also thought to be important. Salt crystals, mainly sodium chloride, are carried by wind from the sea or from ground efflorescences, form solutions when any snow melts and later recrystallize, helping to wedge-out rock crystals. The process has been described for desert areas by R. U. Cooke and I. J. Smalley (1968). Selby, unlike Cailleux, maintains that freeze-thaw action is negligible in the McMurdo Sound area compared with salt weathering, and support for this view is given by C. A. Cotton and A. T.

Wilson (1971). They argue that salt weathering is responsible for widespread 'paring-down' of the landscape in frigid-arid areas, even capable of destroying former glacial features. Slopes of broken debris accumulate below steep rock faces; the debris is further comminuted by salt weathering and when small enough is removed by wind action.

Further support for the activity of chemical weathering in some arctic areas is provided by Washburn (1969) in north-east Greenland. Honeycombed surfaces are again attributed to it, together with removal of fine debris by wind. Chemical coatings are common—oxide rinds on boulders, carbonate coatings on seasonally-thawed ground and under boulders, related to seasonal evaporation, desert varnish and case hardening are all noted. Granular disintegration is particularly evident—some sites yielded over 100 g in 2 years, but it is not clear how much is due to mechanical separation of hydro-thermally-altered rock; the separation process may be frost-wedging or hydration, and the relative contributions of mechanical and chemical weathering cannot therefore be disentangled.

The solution of limestone in arctic environments has been a matter of some controversy. J. Corbel (1959) maintained that the increased solubility of carbon dioxide in low-temperature meltwater would enhance solution rates. On the other hand, it is widely recognized that the absence or sparsity of vegetation and soil in such regions will greatly restrict the availability of organic acids that are so significant in the limestone solution of temperate and tropical regions. D. I. Smith (1972) presents evidence from a limestone area of the Canadian Arctic that does not support Corbel's view. Measured concentrations of calcium and magnesium hardness for snow-melt, pools and streams were found to be far less than for comparable situations in lower latitudes, clearly demonstrating inferior rates of solution.

2 Stream action and valley development in the periglacial zone

River flow in the periglacial zone is characterized by irregularity and sudden fluctuations. Prolonged periods of sub-zero temperatures will result in complete cessation of flow and therefore of all fluvial activity, but during spring or early summer thaw, river ice suddenly begins to break up and may release floods of catastrophic dimensions for a brief period, when considerable erosion and movement of debris occurs. Rivers with such a régime may develop braided sections and aggrade their beds with the debris moved by the spring flood.

A good example is provided by the Mecham River, near Resolute, N.W.T. which has been studied over a period of some years, first by F. A. Cook (1960, 1967) and later by S. B. McCann et al. (1972). The area is underlain by continuous permafrost and mean monthly air temperatures only rise above zero for 2–3 months of the year. Mean annual precipitation at Resolute is only 135 mm, but more than half of this falls as snow (0·75 m). Cook describes how 80–90 per cent of the annual flow of the Mecham River was concentrated in a 10-day period, when the river was wild and impassable, but for the rest of the year it carried little sediment and could be crossed practically anywhere. Table 1.3 presents data collected by McCann et al. for the 1970 flow season; the Mecham River is here compared with a smaller catchment (Jason's Creek, 2·3 km², Plate II) on neighbouring Devon Island. Following a period of snow saturation with

Plate II
Arctic stream valley partially infilled with snow. Fluvial activity is limited except in the short snow-melt season. (C.E.)

meltwater up to 25–26 June, runoff commences and rapidly builds up to the snowmelt flood in the first half of July. At this time, extensive movement of bedload occurs; sometimes at peak velocities (up to 4 m/s), the whole bed is in motion. Suspended sediment loads are also highly variable but generally similar to those of lower-latitude streams, as would be expected, but solute concentrations are lower. Another example of stream activity in a periglacial area is the study of the Colville River, Alaska, by L. Arnborg *et al.* (1967) but, with these notable exceptions, hydrological data are still comparatively scarce in the Arctic. What data there are, however, suggest that fluvial activity has often been underestimated in periglacial regions.

Ideas that fluvial action is relatively unimportant in the periglacial zone have unfortunately been common in the past. The data presented for the Mecham River do not

Table 1.3   Events of the 1970 flow season for two Arctic rivers (McCann *et al.*, 1972)

	River Mecham	Jason's Creek
First day of streamflow	25 June	26 June
First major flood peak	2 July	2 July
Main period of flow	2–16 July	2–17 July
Maximum discharge	8 July (26·6 m³/s)	2 July (1·42 m³/s)
Maximum recorded suspended sediment concentration	2 July (571 ppm)	10 July (291 ppm)
Maximum recorded solute concentration	5 Aug (90 mg/1)	23 July (102 mg/1)

support this view. The pattern of concentrated activity has far greater erosive and transporting potential than a régime in which river flow is evenly distributed through the year. Moreover, it must be borne in mind that the area around Resolute has a typical continental arctic climate with low precipitation; in maritime arctic areas, the geomorphological activity of rivers will be even greater. J. Büdel (1972) illustrates this point in Spitsbergen. He emphasizes that ground ice has already broken the rocks and prepared them for fluvial action. Rivers do not need, therefore, to carry out new erosive work but merely to melt the *Eisrinde,* the upper frozen and highly shattered layer of the permafrost, and transport the debris. Thus they can deepen their beds rapidly. Büdel claims downcutting rates of the order of 1–3 m/1000 years over the last 10,000 years in Spitsbergen, and holds that similar conditions prevailed over wide areas in the Pleistocene.

2.1   *Valley forms*

Stream erosion in the periglacial zone, as in other parts of the world, produces V-shaped forms. Often, however, the floors of valleys are occupied by valley trains of coarse material (see, for instance, S. Rudberg's (1969) description of fluvial landforms on Axel Heiberg Island), representing the prolific debris broken by frost action and ground ice and waiting to be transported by the summer meltwater flood.

(a) Asymmetrical valleys   Geomorphologists have long sought for some connection between the occurrence of valley asymmetry and the extent of the periglacial zone.

H. Poser (1947), for example, thought that valley asymmetry of non-structural origin
was peculiar to the latter and that such valley forms might even afford a useful clue
to the former extent of frozen ground. The problem is to discover a satisfactory reason
or group of reasons why asymmetrical valleys should be related to cold conditions.
In this discussion we shall not be concerned with asymmetry of structural origin, though
it is often far from easy to assess and to separate the influence of rock structure.

Valley asymmetry has been investigated in widely differing periglacial conditions
and it is abundantly clear that the steeper slope of the asymmetrical pair may face
in any direction, though preferred orientations are often characteristic of individual
regions. In western Europe, the steeper slope most commonly faces west or south-west
(Fig. 1.4). This is noted by Büdel (1953) as a general characteristic, by A. Gloriod
and J. Tricart (1952) in the Pas de Calais of northern France, by J. Alexandre (1958)
in the Ardennes, by K. Helbig (1965) in eastern Austria and southern Germany, by

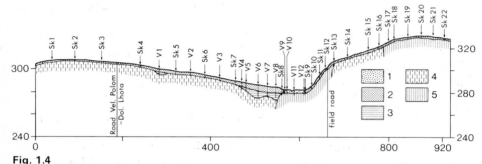

Fig. 1.4

Cross-section of an asymmetrical valley in the Nízký Jeseník mountains, Czechoslovakia. (T. Czudek
and J. Demek, *Institute of British Geographers, Transactions* 1971) Scales in metres
1 Flood-plain deposits
2 Loam and clay with Culm rock fragments
3 Clay from *in situ* Tertiary weathering
4 Strongly weathered Culm shales and greywackes
5 Slightly weathered Culm shales and greywackes

T. Czudek and J. Demek (1971) in western Czechoslovakia and by Czudek (1973)
in Moravia. In the Chiltern Hills of England, C. D. Ollier and A. J. Thomasson (1957)
find a dominant south or south-west aspect for the steeper side, as does H. M. French
(1971) for the far-distant area of Banks Island in Arctic Canada. In the Mackenzie
River area of arctic Canada, B. A. Kennedy and M. A. Melton (1972) note that no
one form of asymmetry dominates: both north- and south-facing slopes may be steeper.
D. R. Currey (1964) has examined asymmetrical valleys in western Alaska, from Cape
Lisburne to the Arctic Circle. Table 1.4 summarizes some of the data obtained. In
this area, wherever streams flow roughly due north or south, the number of valleys
with steeper left-bank slopes is approximately equal to the number with steeper right-
bank slopes, but as the direction of stream flow departs from a north-south azimuth,
valley asymmetry becomes strongly marked, with a definite preponderance of steeper
north-facing sides.

With such variety of asymmetry from region to region and also within regions, it
seems unlikely that there can be one explanation or one hypothesis applicable to every

Table 1.4   Analysis of asymmetrical valleys in part of Alaska (Currey, 1964)

Azimuth of stream flow	Number of valleys with steeper left-bank slopes	Number of valleys with steeper right-bank slopes	Valleys showing little or no asymmetry	Totals
1–180°	8	24	15	47
181–360°	104	15	66	185
	112	39	81	232

area, but there are certain sets of conditions or processes peculiar to the periglacial zone which contribute to valley asymmetry:

1 Effect of insolation: north-facing slopes will remain in shadow for longer; thus they may remain frozen for longer in permafrost areas and retain any initial steepness for longer. South-facing slopes will more readily thaw and experience gelifluction, creep and downhill transport by any meltwater.

2 Effect of permafrost: although it has never been established that asymmetrical valleys are restricted to areas of present or former permafrost, permafrost plays an important part in slope development, stabilizing slopes of unconsolidated materials at angles steeper than could be maintained in the thawed state.

3 Effect of prevailing winds: snow will accumulate mainly, or to greater depths, on lee slopes. In summer, snow patches will thus tend to survive on the latter and, by providing meltwater, encourage downhill movement of debris. Chapter 5 provides a fuller discussion of the effects of nivation. It is far from clear whether the presence of snow patches leads to reduction of overall slope angle or not.

4 Effect of stream action: basal undercutting of a valley side is important in steepening that side and removing any debris from its foot. On permafrost, thermo-erosive action of the relatively warm stream water will be a part of the process of basal undercutting. Greater debris supply on one valley slope will have the effect of tending to push the stream channel over and cause undercutting of the other slope.

5 Effect of contrasts in plant cover: the influence of vegetation may be extremely important in affecting ground freezing and thawing, as L. W. Price (1971, 1972a) has demonstrated. Four slopes with different aspects (north, east, south-east and south-west) but with similar gradients and elevations were studied in the Ruby Range, Alaska. Vegetation was best developed on the south-east facing slopes, less on the east- and south-west facing slopes, and least on the north-facing slope. On the south-east facing slope, paradoxically, the active layer was shallowest because the thicker vegetation insulated it best. Conversely, against expectation, the north-facing slope thawed more quickly and deeply because of the sparse plant cover and the higher conductivity of bare soil, in spite of the much lower angle of the sun's rays.

6 Effect of animals: although of local importance only, activity of burrowing animals should not be overlooked. Price (1972b) showed how colonies of arctic ground squirrels almost exclusively favour south-east facing slopes on the Ruby Range, Alaska, where vegetation and gelifluction lobes are well developed and where insolation is greatest. Burrowing is estimated to be responsible for excavating 20 t/ha

annually over the area of its occurrence. The squirrels also push material downslope, undermine boulders and cause slope collapse by tunnelling.

It is clear that the possible combinations of any or all of these factors, each of which may vary in its level of intensity, are so numerous that there is no simple explanation to periglacial valley asymmetry. In our present state of knowledge, as Washburn (1973) points out, it is dangerous to make inferences about the extent of the Pleistocene periglacial zone or of Pleistocene permafrost from the distribution of present-day asymmetrical valleys.

(b) Dells   This term is applied to the small shallow valleys with concave cross-profiles, now usually dry except during snowmelt or occasional rainstorms, which characterize some Pleistocene periglacial areas (Büdel, 1953; Czudek and Demek, 1971). Under conditions of impermeable permafrost, sparse vegetation and melting snow, surface runoff was encouraged, as C. Reid deduced as long ago as 1887. The floors of dells are usually infilled with angular gelifluction rubble. There are, of course, many possible non-periglacial origins for dry valleys in present-day landscapes, and to distinguish dells from these other types is fraught with difficulty. Like asymmetrical valleys, great caution should be exercised before using such vaguely defined forms to delimit Pleistocene periglacial influence.

The formation of dells under modern arctic conditions has been observed by Czudek and Demek (1973) in eastern Siberia. Here, disturbance of the thermal equilibrium induces selective thawing of ice wedges (see Chapter 2). Small channels develop by thermo-erosion, and material moves downslope into them by surface wash and gelifluction as fast as it is removed along the channel, thus preserving a shallow cross-section.

(c) Vallons de gélivation   M. Boyé (1950) reported the occurrence in Greenland of valleys ('ravins de gélivation') believed to be formed not by stream action but by the widening of lines of structural weakness by frost action and the evacuation of this debris by gelifluction. A similar explanation for some valleys near Knob Lake, central Labrador, was proposed by C. R. Twidale (1956). These valleys are cut in the side of an argillite ridge, some ending abruptly in vertical rock walls but opening out in the other direction on to a shale lowland. In cross section, the valleys show flanking talus slopes beneath rocky scarps; the largest valley is about 10 m deep. Protalus ramparts (see p. 140) mark their mouths and were considered by Twidale to mark the soliflual movement of material down-valley to their exits. J. T. Andrews (1963b) has criticized some of Twidale's findings. Although he accepts that the valleys may have been initiated by joint-guided frost shattering, he proposes a pre-Wisconsin rather than a post-glacial age for this, and produces evidence to suggest that the valleys have been much modified by subglacial meltwater erosion during the Wisconsin glaciation—a large ice-emplaced erratic at the mouth of one valley bears witness to an ice-cover *since* the valley was formed. Detailed mapping by W. Barr (1969) supports the view that these valleys are structurally-guided subglacial meltwater channels. Andrews deplores the tendency to assume that freeze-thaw action is necessarily impor-

tant in a region of cold climate; as already noted, conditions in many high-latitude arctic regions today are distinctly unfavourable for freeze-thaw action, with temperatures continuously below freezing in the winter and continuously above freezing in the summer.

Boyé's hypothesis remains a possible one for the formation of certain structurally-guided small-scale valleys in present or former periglacial regions, but it should be applied with caution and with alternative modes of origin carefully considered.

3 Wind action in the periglacial zone

Wind action is responsible for one of the most widespread and bulky deposits of the periglacial zone, namely loess, together with deposits of coarser materials forming coversands and dunes. These deposits and their associated landforms will be examined in detail in Chapter 7, as will the rather rarer examples of periglacial wind erosion. The more general role of the wind in causing snow drifting on lee slopes and in gullies, and in keeping other areas swept clean of snow, must not be forgotten. Areas without snow become colder by exposure and lack possible moisture from snowmelt (French, 1972b). Thus the wind exerts an important indirect influence on gelifluction (Chapter 4) and on nivation (Chapter 5), and also on some coastal features which will be considered next.

4 Coastal and lake-shore features in the periglacial zone

Arctic coasts are dominated for part or all of the year by the effects of sea ice. Sea ice limits wave generation and wave action, so that these are typically low-energy coasts (C. A. M. King, 1969). The existence of permafrost in the beach material further inhibits wave action. The permafrost table represents the effective lower limit of beach processes and also prevents percolation so that, unlike coasts outside the periglacial zone, the volumes and velocities of swash and backwash may be roughly equal (E. H. Owens and S. B. McCann, 1970). Although permafrost inhibits the free movement of beach material at more than shallow depth, it does not prevent recession of the coastline by thermo-erosion. R. I. Lewellen (1970) estimates that thermo-erosion rates on the Beaufort Sea coast of northern Alaska approach 10 m/year. Because of the low wave energy, the beach material is relatively unsorted compared with non-arctic coasts, and the material itself is frequently angular or sub-angular, with numerous frost-shattered stones. A cover of frozen sea-spray and swash may develop in the autumn and prevent movement of beach material until wave action itself comes to a halt as the sea freezes in winter. On Devon Island (Canada, latitude 75°N), Owens and McCann (1970) note that the period of effective wave action is as little as 8 weeks in a year (Plate III).

The most distinctive features of arctic coasts are associated with ice-push. The power of pack ice driven on-shore by wind (or to a lesser extent by sea currents) may be very considerable. Even in the Gulf of Finland with a width of only 80–120 km, pack ice is capable of moving boulders up to 15 m³ in size, of pushing up 'block walls' 1–2 m high, with smaller blocks packed between larger boulders, of collecting

Plate III
Coast of Devon Island, Canadian Arctic. Sea ice, breaking up in July or August, disturbs the beach
sediments by ice-push or by melting of ice blocks enclosed in the sediment. (C.E.)

smaller blocks around larger unmoved blocks (H. Mansikkaniemi, 1970) and of producing striations on the shore platform. On more exposed arctic coasts, the push of
sea ice is known to affect shores for tens of metres inland and to form ridges and
ramparts several metres high, though permafrost prevents the disturbance from extending to more than shallow depths. As long ago as 1919, E. K. Leffingwell reported from the Canning River district of Alaska how sea ice was able to override
the shore for 6 m or more and to pile up ridges 2 m high.

McCann and R. J. Carlisle (1972) describe how the sea ice becomes frozen in winter
to the beaches of Radstock Bay, Devon Island. The tidal range here is about 2·5 m,
and this allows a layer of 'fast ice' to develop on the beach, known as an ice foot. In
non-tidal environments, the frozen swash layer with interbedded sediment is termed
a kaimoo. Pressure transmitted from the sea ice causes buckling of both sediments and
the beach-fast ice. In spring, the sea ice will part company and leave the ice foot behind,
continuing to protect the beach from wave action for a time. Thereafter, a series of
features resulting from the gradual melting of the ice foot may be formed. Ice-cored
mounds of beach sediment, produced when storms buckle and pile up beach ice and
sediment, will gradually melt in summer to leave behind small heaps of gravel. Sea
ice buried in the beach sediment will slowly melt away and leave miniature kettle-holes.

The assemblage of ice-push ramparts, pits and mounds is a common characteristic of arctic coasts; their degree of development depends on the relative exposure of the coast and the duration of sea freezing. J. D. Hume and M. Schalk (1964, 1967) describe the effects on the more open beaches of north Alaska.

Similar but smaller-scale features are associated with ice on lake shores (J.-C. Dionne and C. Laverdière, 1972). As smaller lakes will freeze over completely, wind may be powerless to shift the ice, and when the ice breaks up in spring, the limited fetch will mean that ice push by wind will be less significant than on open sea coasts. However, another possible mechanism of ice push has been suggested for lakes. J. Hult (1968) points out that lake ice will shrink with a fall of temperature below 0°C. Cracks will develop in the ice cover anchored to the shore; water will fill the cracks and freeze. With rise of temperature, the ice cover will expand and thrust against the shore. For this process to be effective, Hult suggests that the lake must be small (<3 km), the ice must be at least 20 cm thick, and the rise in temperature must be considerable and prolonged, possibly at least 0·5°C/hour over a period of 12 hours.

5 Conclusion

This chapter has introduced the subject of periglacial geomorphology and considered certain aspects of the periglacial environment. Weathering processes and the parts played by running water, wind and marine action have been outlined. The following chapters will deal in more detail with the most distinctive features and processes connected with the periglacial zone: frozen ground phenomena, patterned ground, mass movements and slope deposits, the action of snow both mobile and stationary, and the action of wind.

6 References

ALEXANDRE, J. (1958), 'Le modelé quaternaire de l'Ardennes Centrale', *Ann. géol. Soc. Belge* **81**, 213–332

ANDERSSON, J. G. (1906), 'Solifluction, a component of sub-aerial denudation', *J. Geol.* **14**, 91–112

ANDREWS, J. T. (1961), 'Vallons de gélivation in central Labrador-Ungava: a reappraisal', *Can. Geogr.* **5**, 1–9

(1963a), 'The analysis of frost-heave data collected by B. H. J. Haywood from Schefferville, Labrador-Ungava', *Can. Geogr.* **7**, 163–73

(1963b), 'So-called "vallons de gélivation" in central Labrador-Ungava', *Biul. Peryglac.* **12**, 137–43

ARNBORG, L., WALKER, H. J. and PEIPPO, J. (1967), 'Suspended load in the Colville River, Alaska, 1962', *Geogr. Annlr* **49**, 131–44

BARR, W. (1969), 'Structurally-controlled fluvioglacial erosion features near Schefferville, Quebec', *Cah. Géogr. Québec* **13**, 295–320

BOYÉ, M. (1950), *Glaciaire et périglaciaire de l'Ata Sund nord-oriental (Groenland)* (Paris; Expéditions polaires françaises, **1**)

BRIDGMAN, P. W. (1912), 'Water in the liquid and five solid forms, under pressure', *Am. Acad. Arts Sci.* **47**, 439–558

BRYAN, K. (1946), 'Cryopedology—the study of frozen ground and intensive frost action with suggestions on nomenclature', *Am. J. Sci.* **244**, 622–42

BÜDEL, J. (1953), 'Die "periglazial"-morphologischen Wirkungen des Eiszeitklimas auf der ganzen Erde', *Erdkunde* **7**, 249–66

—— (1959), 'Periodische und episodische Solifluktion im Rahmen der Klimatischen Solifluktionstypen', *Erdkunde* **13**, 297–314

—— (1972), 'Typen der Talbildung in verschiedenen klimamorphologischen Zonen', *Z. Geomorph., Suppl. Bd* **14**, 1–20

CAILLEUX, A. (1962), *Études de géologie au détroit de McMurdo (Antarctique)*, *Com. natn. fr. Rech. Antarct.* **1**, 41 pp.

—— (1968), 'Periglacial of McMurdo Strait (Antarctica)', *Biul. Peryglac.* **17**, 57–90

CAIRNES, D. D. (1912), 'Differential erosion and equiplanation in portions of Yukon and Alaska', *Bull. geol. Soc. Am.* **23**, 333–48

CONNELL, D. C. and TOMBS, J. M. C. (1971), 'The crystallization pressure of ice—a simple experiment', *J. Glaciol.* **10**, 312–15

COOK, F. A. (1960), 'Periglacial-geomorphological investigations at Resolute, 1959', *Arctic* **13**, 132–5

—— (1967), 'Fluvial processes in the High Arctic', *Geogr. Bull.* **9**, 262–8

COOK, F. A. and RAICHE, V. G. (1962), 'Freeze-thaw cycles at Resolute, N.W.T.', *Geogr. Bull.* **18**, 64–78

COOKE, R. U. and SMALLEY, I. J. (1968), 'Salt weathering in deserts', *Nature, Lond.* **220**, 1226–7

CORBEL, J. (1959), 'Érosion en terrain calcaire: vitesse d'érosion morphologique', *Annls Géogr.* **366**, 97–120; and 'Vitesse de l'érosion', *Z. Geomorph.* **3**, 1–28

COTTON, C. A. and WILSON, A. T. (1971), 'Pared-down landscapes in Antarctica', *Earth Sci. J.* **5**, 1–15

CURREY, D. R. (1964), 'A preliminary study of valley asymmetry in the Ogotoruk Creek area, north-western Alaska', *Arctic* **17**, 84-98

CZEPPE, Z. (1960), 'Thermic differentiation of the active layer and its influence upon the frost heave in periglacial regions (Spitsbergen)', *Bull. Acad. pol. Sci. classe III*, **8**, 149–52

—— (1964), 'Exfoliation in a periglacial climate', *Geogr. Polonica* **2**, 5–10

CZUDEK, T. (1973), 'Die Talasymmetrie im Nordteil der Moravská Brána', *Acta Sci. Nat. Brno* **7**, 1–48

CZUDEK, T. and DEMEK, J. (1971), 'Pleistocene cryoplanation in the Česká vysočina highlands, Czechoslovakia', *Trans. Inst. Br. Geogr.* **52**, 95–112

—— (1973), 'Die Reliefentwicklung während der Dauerfrostbodendegradation', *Rozpr. čsl. Akad. Ved.* **83**, 69 pp.

DAVIES, W. E. (1961), 'Surface features of permafrost in arid areas', *Folia Geogr. Danica* **9**, 48–56

DIONNE, J.C. and LAVERDIÈRE, C. (1972), 'Ice-formed beach features from Lake St. Jean, Quebec', *Can. J. Earth Sci.* **9**, 979–90

DOUGLAS, G. R. (1972), 'Processes of weathering, and some properties of the Tertiary basalts of Co. Antrim, northern Ireland', Unpubl. Ph.D. thesis, Queen's University of Belfast

DYLIKOWA, A. and OLCHOWIK-KOLASINSKA, J. (1956), 'Processes and structures in the active zone of perennially frozen ground', *Biul. Peryglac.* **3**, 119–24

EAKIN, H. M. (1916), 'The Yukon-Koyukuk region, Alaska', *U.S.geol. Surv. Bull.* **631**, 1–88

EVERETT, D. H. (1961), 'The thermodynamics of frost damage to porous solids', *Trans. Faraday Soc.* **57**, 1541–51

FALCONER, A. (1969), 'Processes acting to produce glacial detritus', *Earth Sci. J.* **3**, 40–3

FRASER, J. K. (1959), 'Freeze-thaw frequencies and mechanical weathering in Canada', *Arctic* **12**, 40–53

FRENCH, H. M. (1971), 'Slope asymmetry of the Beaufort Plain, north-west Banks Island, N.W.T., Canada', *Can. J. Earth Sci.* **8**, 717–31

— (1972a), 'Asymmetrical slope development in the Chiltern Hills', *Biul. Peryglac.* **21**, 51–73

— (1972b), 'The role of wind in periglacial environments, with special reference to north-west Banks Island, western Canadian Arctic' in *International Geography 1972* (Montreal) **1**, 82–4

GEIKIE, J. (1874), *The Great Ice Age* (London, 1st Ed.)

GLORIOD, A. and TRICART, J. (1952), 'Étude statistique des vallées asymmétriques de la feuille St Pol au 1:500,000', *Rev. Géom. dyn.* **3**, 88–98

GRAWE, O. R. (1936), 'Ice as an agent of rock weathering', *J. Geol.* **44**, 173–82

GUILLIEN, Y. and LAUTRIDOU, J.-P. (1970), 'Recherches de gélifraction expérimentale du Centre de Géomorphologie', *Bull. Centre Géomorph. Caen*, **5**

HAMELIN, L.-E. (1961), 'Périglaciaire du Canada: idées nouvelles et perspectives globales', *Cah. Géogr. Québec* **10**, 141–203

HAMELIN, L.-E. and COOK, F. A. (1967), *Illustrated glossary of periglacial phenomena* (Quebec)

HELBIG, K. (1965), 'Asymmetrische Eiszeittäler in Süddeutschland und Ostösterreich', *Würzb. Geogr. Arb.* **14**, 1–108

HEWITT, K. (1968), 'The freeze-thaw environment of the Karakoram Himalaya', *Can. Geogr.* **12**, 85–98

HÖGBOM, B. (1914), 'Über die geologische Bedeutung des Frostes', *Bull geol. Instn Univ. Upsala* **12**, 257–389

HÖLLERMANN, P. W. (1967), 'Zur Verbreitung rezenter periglazialer Kleinformen in den Pyrenäen und Ostalpen (mit Ergänzungen aus dem Apennin und dem Französischen Zentralplateau)', *Göttinger Geogr. Abh.* **40**, 1–198

HULT, J. (1968), 'Some aspects of the shore formations on Lake Lylykkäänjärvi, Finland', *Fennia* **97**, no. 5, 22 pp.

HUME, J. D. and SCHALK, M. (1964), 'The effects of ice push on Arctic beaches', *Am. J. Sci.* **262**, 267–73

— (1967), 'Shoreline processes near Barrow, Alaska. A comparison of the normal and the catastrophic', *Arctic* **20**, 86–103

JAHN, A. (1954), 'Walery Lozinski's merits for the advancement of periglacial studies', *Biul. Peryglac.* **1**, 117–24

KENNEDY, B. A. (1969), 'Periglacial morphometry' in *Water, Earth and Man* (ed. R. J. CHORLEY), 381–8

KENNEDY, B. A. and MELTON, M. A. (1972), 'Valley asymmetry and slope forms of a permafrost area in the Northwest Territories, Canada', *Inst. Br. Geogr. Spec. Publ.* **5**, 107–21

KESSLER, D. W. *et al.* (1940), 'Physical, mineralogical and durability studies on the building and monumental granites of the United States', *Natn. Bur. Stand. Res. Pap.* **1320**, 161–206

KING, C. A. M. (1969), 'Some Arctic coastal features around Foxe Basin and in east Baffin Island, N.W.T., Canada', *Geogr. Annlr* **51**A, 207–18

LEFFINGWELL, E. K. (1919), 'The Canning River region, northern Alaska', *U.S. geol. Surv. Prof. Pap.* **109**, 1–251

LEWELLEN, R. I. (1970), *Permafrost erosion along the Beaufort Sea coast* (Denver, Col.), 25 pp.

LINTON, D. L. (1969), 'The abandonment of the term "periglacial"' in *Paleo-ecology of Africa and of the surrounding islands and Antarctica* (ed. E. M. VAN ZINDEREN BAKKER) (Cape Town), **5**, 65–70

LOZINSKI, W. (1912), 'Die periglaziale Fazies der mechanischen Verwitterung', *C. r. 11th int. geol. Congr. (Stockholm, 1910)*, **2**, 1039–53

MCCANN, S. B. and CARLISLE, R. J. (1972), 'The nature of the ice-foot on the beaches of Radstock Bay, south-west Devon Island, N.W.T., Canada, in the spring and summer of 1970', *Inst. Br. Geogr. Spec. Publ.* **5**, 175–86

MCCANN, S. B., HOWARTH, P. J. and COGLEY, J. G. (1972), 'Fluvial processes in a periglacial environment: Queen Elizabeth Islands, N.W.T., Canada', *Trans. Inst. Br. Geogr.* **55**, 69–82

MANSIKKANIEMI, H. (1970), 'Ice-push action on sea shores, south-eastern Finland', *Publs Inst. Geogr. Univ. Turkuensis* **50**, 1–30

MATTHES, F. E. (1900), 'Glacial sculpture of the Big Horn Mountains, Wyoming', *U.S. geol. Surv. 21st A. Rep.* (1899–1900), 167–90

OLLIER, C. D. and THOMASSON, A. J. (1957), 'Asymmetrical valleys of the Chiltern Hills', *Geogrl J.* **123**, 71–80

OWENS, E. H. and MCCANN, S. B. (1970), 'The role of ice in the Arctic beach environment, with special reference to Cape Ricketts, south-west Devon Island, Northwest Territories, Canada', *Am. J. Sci.* **268**, 397–414

PELTIER, L. C. (1950), 'The geographic cycle in periglacial regions as it is related to climatic geomorphology', *Ann. Ass. Am. Geogr.* **40**, 214–36

POSER, H. (1947), 'Dauerfrostboden und Temperaturverhältnisse während der Würmeiszeit im nacht vereisten Mittel- und Westeuropa', *Naturwissenschaften, Berlin* **34**, 10–18

— (1948), 'Boden- und Klimaverhältnisse in Mittel- und Westeuropa ẅhrend der Würmeiszeit', *Erdkunde* **2**, 53-68

POTTS, A. S. (1970), 'Frost action in rocks: some experimental data', *Trans. Inst. Br. Geogr.* **49**, 109–24

PRICE, L. W. (1971), 'Vegetation, microtopography and depth of active layer on different exposures in subarctic alpine tundra', *Ecology* **52**, 638–47

(1972a), ibid., *Icefield Ranges Res. Proj. scient. Results* **3**, 211–20

(1972b), 'Geomorphic effect of the arctic ground squirrel in an alpine environment', *Icefield Ranges Res. Proj. scient. Results* **3**, 255–9

RAPP, A. (1960), 'Recent development of mountain slopes in Kärkevagge and surroundings, northern Scandinavia', *Geogr. Annlr.* **42**, 65–200

REID, C. (1887), 'On the origin of the dry chalk valleys and of the coombe rock', *Q. J. geol. Soc. Lond.* **43**, 364–73

ROGNON, P., CUSSENOT-CURIEN, M., SEYER, C. and WEISROCK, A. (1967), 'Remarques sur le comportement des grès et granites vosgiens sous climat froid', *Revue Géogr. de l'Est* **7**, 403–18

RUDBERG, S. (1969), 'Distribution of small-scale periglacial and glacial geomorphological features on Axel Heiberg Island, N.W.T., Canada', in *The periglacial environment* (ed. T. L. PÉWÉ, Montreal), 129–59

(1972), 'Periglacial zonation: a discussion', *Göttinger Geogr. Abh.* (Hans Poser Festschrift) **60**, 221–33

SELBY, M. J. (1972), 'Antarctic tors', *Z. Geomorph., Suppl.* **13**, 73–86

SMITH, D. I. (1972), 'The solution of limestone in an Arctic environment', *Inst. Br. Geogr. Spec. Publ.* **5**, 187–200

SMITH, H. T. U. (1949), 'Physical effects of Pleistocene climatic changes in non-glaciated areas: eolian phenomena, frost action and stream terracing', *Bull. geol. Soc. Am.* **60**, 1485–516

STINGL, H. (1969), 'Ein periglazialmorphologisches Nord-Süd-Profil durch die Ostalpen', *Göttinger Geogr. Abh.* **40**, 1–115

STODDART, D. R. (1969), 'Climatic geomorphology: review and re-assessment', *Progr. Geogr.* **1**, 160–222

TABER, S. (1943), 'Perennially frozen ground in Alaska: its origin and history', *Bull. geol. Soc. Am.* **54**, 1433–548

TAYLOR, G. (1922), '*The physiography of the McMurdo Sound and Granite Harbour region*, British Antarctic (Terra Nova) Expedition 1910–1913 (London), 246 pp.

TRICART, J. (1956), 'Étude expérimentale du problème de la gélivation', *Biul. Peryglac.* **4**, 285–318

(1967), *Le modelé des régions périglaciaires* (Paris)

(1969), *Geomorphology of cold environments* (trans. E. WATSON)

TROLL, C. (1944), 'Strukturböden, Solifluktion, und Frostklimate der Erde', *Geol. Rdsch.* **34**, 545–694 (English translation, *Snow Ice Permafrost Res. Establ.*)

TWIDALE, C. R. (1956), 'Vallons de gélivation dans le centre du Labrador', *Revue Géomorph. dyn.* **7**, 17–23

WASHBURN, A. L. (1969), 'Weathering, frost action and patterned ground in the Mesters Vig district, north-east Greenland', *Meddr Grønland* **176** (4), 303 pp.

(1973), *Periglacial processes and environments* (London)

WILSON, L. (1969), 'Les relations entre les processus géomorphologiques et le climat moderne comme méthode de paleoclimatologie', *Rev. Géogr. phys. Géol. dyn.* **11**, 303–14

WIMAN, S. (1963), 'A preliminary study of experimental frost weathering', *Geogr. Annlr* **45**, 113–21

ZEUNER, F. E. (1945), *The Pleistocene period*

2

Frozen ground phenomena

...banks, which are about six feet above the surface of the water, display a face of solid ice, intermixed with veins of black earth, and as the heat of the sun melts the ice, the trees frequently fall into the river. (SIR ALEXANDER MACKENZIE, *Voyages to the Arctic*, 1789)

As early as the sixteenth century, explorers were bringing back reports of frozen ground from Arctic regions, and in the seventeenth century, news of the discovery of the frozen carcases of mammoth and woolly rhinoceros in Siberia reached Europe. The first intelligent description of ground ice is ascribed to M. F. Adams, reporting from the Lena delta in 1806, while in North America, O. von Kotzebue's expedition of 1816 recorded ground ice at Elephant Point, Alaska. The earliest detailed studies of frozen ground were made by A. T. von Middendorff, who collected data from wells dug in search of water in Yakutsk, Siberia (G. B. Cressey, 1939). Interest in frozen ground was greatly stimulated during the construction of sections of the Trans-Siberian Railway, for entirely new problems of engineering were presented by such terrain, while in Alaska and the Yukon, the discovery and exploitation of alluvial gold also brought about closer acquaintance with the characteristics of frozen ground. A landmark in the growing literature on frozen ground was the publication in 1919 of E. K. Leffingwell's monograph on the Canning River region in Alaska, which included a major section on such topics as the depths and temperature gradients of the frozen ground, the forms, age, and origins of ground ice, and on the related landforms. Since then, the investigation of frozen ground has accelerated to such an extent that its study can now almost be regarded as a separate scientific discipline, claiming attention not only from geologists and geographers but also from geophysicists, engineers and all those concerned with land utilization in arctic and sub-arctic regions. Particularly rapid advances were made during World War II for military reasons, when the US Army established SIPRE (Snow, Ice and Permafrost Research Establishment), later renamed CRREL (Cold Regions Research and Engineering Laboratory). US Army CRREL produced a useful report on permafrost in 1966 (S. R. Stearns, 1966). More recently still, discoveries of oil at Prudhoe Bay and elsewhere on the North Slope of Alaska have focused attention on the permafrost and its problems, as has the proposal to build a 1300-km pipeline

southwards to ice-free tidewater. The greatest contributions to knowledge of the permafrost over the last 30 years or so have come from Soviet scientists. In the USSR, permafrost underlies an area two and a half times that of Canada, and its maximum thickness
may also be twice as great. Major cities and industrial enterprises have been successfully
established in the Soviet Arctic, showing that the permafrost is not the insuperable
obstacle to economic progress as was once thought.

1 The present-day occurrence of permafrost

S. W. Muller (1947) proposed the name 'Permafrost' for a 'thickness of soil or other
superficial deposit, or even of bedrock, at a variable depth ... in which a temperature
below freezing has existed continually for a long time' (p. 3). He distinguished between
'dry permafrost', in which moisture was lacking or insufficient to allow interstitial ice
to form and act as a cementing material, and 'frozen ground' which contained water
mostly in the form of ice. In permafrost, frozen and unfrozen water may exist in equilibrium, depending on local variations of temperature, pressure, impurities in the water,
and the grain size of the material. The term 'permafrost' has often been criticized (K.
Bryan, 1946, suggested its replacement by 'pergelisol') but its usage is now well
established as a useful and easily intelligible contraction of 'Permanently frozen
ground'.* Cold glaciers are 'permafrost' by Muller's definition, but conventionally they
are excluded from the term. Permafrost may be ice-free at up to several degrees below
zero if the contained water is saline or if the soil contains much clay.

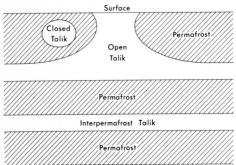

Fig. 2.1
Idealized diagram showing various possible
forms of talik in permafrost (J. Demek, *Biul.
Peryglac.*, 1969)

Above permafrost there may be a zone in which temperature conditions vary seasonally. This is the 'active layer' (Bryan's mollisol) thawing out when temperatures rise
sufficiently, and refreezing in winter or in cold spells (or even partly at night time)
to depths from a few millimetres to about 3 m depending on the climate. In this layer,
plants are rooted, and movements can readily occur in the thawed state. The thickness
of the active layer depends on the air temperature régime, the degree of exposure to
insolation (slope aspect is important), the conductivity of the soil, and the degree of
insulation which might be provided by a cover of snow or vegetation. The active layer
may rest directly on the upper surface of permafrost (the 'permafrost table'), but sometimes, an unfrozen layer may intervene (see, for example, J. D. Ives, 1960). Unfrozen
layers are also known within permafrost; they are termed 'taliks' (Fig. 2.1). They may

* Strictly, one should speak of *perennially* rather than *permanently* frozen ground.

reflect former climatic changes—the relations between permafrost and climate will be considered later—or they may represent aquifers containing water highly mineralized or under pressure, for instance. Beneath all permafrost, unfrozen ground exists, for permafrost is simply the result of negative heat balance at the earth's surface, either now or in the past, replacing the normal positive temperature gradient caused by out-flow of heat from the earth's interior.

Surface features have an important effect on the local thickness and distribution of permafrost (O. J. Ferrians *et al.*, 1969; Fig. 2.2). Surface water effectively shuts off the penetration of sub-zero temperatures from the surface downwards. Thus permafrost does not generally extend beyond sea coasts (but see pp. 29–30), nor beneath large deep lakes. The latter will give rise to unfrozen 'chimneys' in the permafrost. The effect of small lakes depends on their depth and whether they freeze or not in winter; small deep lakes remaining unfrozen will be underlain by a thawed 'basin' in the upper sur-face of the permafrost, as Fig. 2.2 shows, and a corresponding upward indentation

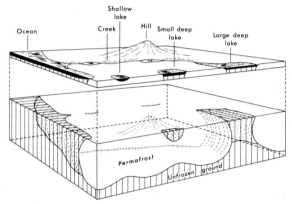

Fig. 2.2
Effect of surface features on the distribution of permafrost in a continuous permafrost area (A. H. Lachenbruch, 1968, *Encyclopaedia of Geomorphology* (ed. R. W. Fairbridge), New York, p. 837)

in the base of the permafrost. Small shallow lakes freezing in winter have only slight effect on the permafrost thickness.

Conventionally, three spatial zones of permafrost occurrence are recognized (L. L. Ray, 1951):

1 *continuous permafrost*, mainly related to areas where the climate now is severe enough to form permafrost; in this zone, the only unfrozen areas are those lying below certain deep and wide lakes, rivers, or beneath the sea;
2 *discontinuous permafrost*, in which small scattered unfrozen areas appear; and
3 *sporadic permafrost*, where small islands of permafrost exist in a generally unfrozen area.

It is likely that many such islands are relics of a former colder climate, and are steadily diminishing in size. Continuous or discontinuous permafrost underlies as much as one-quarter of the earth's land areas, including 47 per cent of the USSR, possibly as much as 50 per cent of Canada and Alaska, and most of Greenland and Antarctica. If,

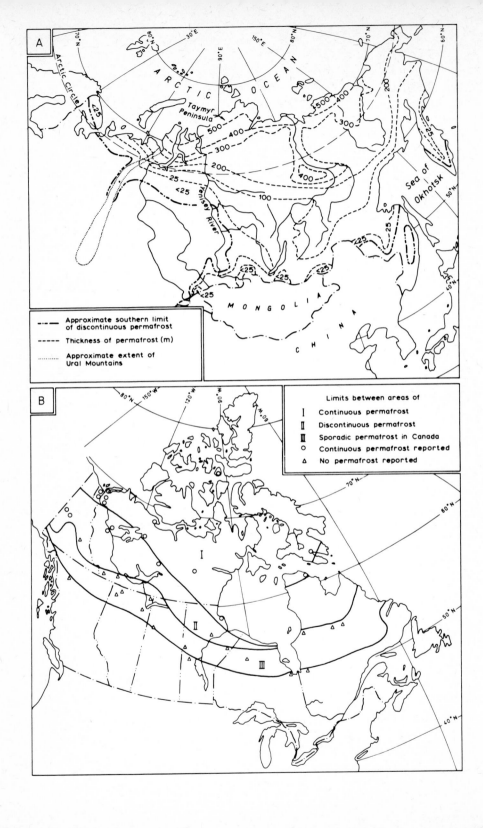

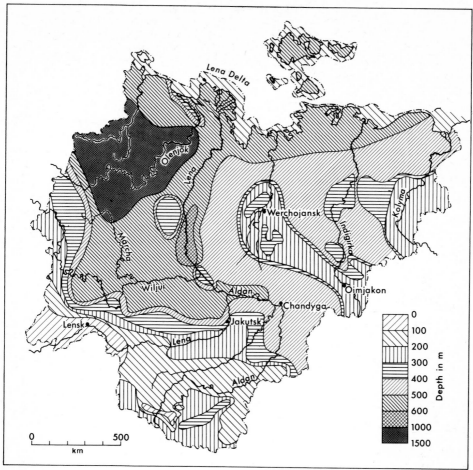

Fig. 2.4
Permafrost in the Yakutian ASSR (P. I. Melnikov, 1970, from 1 : 5 million map *Schematitscheskaja geokriologitscheskaja Karta Jakutskoi ASSR*, Moscow).

however, present-day glacierized areas are excluded, for we know little about the extent of permafrost beneath these, the overall proportion is about 14 per cent.

Fig. 2.3A shows the distribution of permafrost in the USSR, where sufficiently de-tailed knowledge is available to plot isopleths of its thickness: no comparable map can yet be drawn for Canada and Alaska. It should be noted how permafrost underlies parts of the Arctic Ocean floor; these submarine areas possess temperatures below 0°C and are therefore permafrost by definition, but they generally contain no ice because of the salinity conditions. Siberia includes the coldest area of the world outside Antarc-tica; at Oymyakon on the Indigirka River (latitude 66°N.), an absolute minimum

Fig. 2.3
A Distribution and thickness of permafrost in the USSR (R. J. E. Brown, *Arctic*, 1960, Arctic Institute of North America);
B Permafrost zones in Canada (R. J. E. Brown, *Geol. Surv. Canada*, 1967)

temperature of $-71°C$ has been recorded (I. P. Gerasimov, 1961). Since this map was prepared, more data have become available, making it necessary to revise the details, though the general patterns of permafrost thicknesses have not altered radically. Fig. 2.4 shows a map compiled for Yakutia in 1970, including the area where the greatest known thicknesses of permafrost in the world have been discovered.

In Fig. 2.3B, the conventional zones of permafrost in Canada are indicated, but the map is far from being a precise representation owing to inadequate data. An indication of this was provided during iron ore company excavations and drillings in central Labrador (Ives, 1960; J. T. Andrews, 1961), for 60-metre thicknesses of permafrost

Table 2.1   Some permafrost thicknesses in the Arctic and comparisons with present mean annual air temperature

		Height (m)	Mean annual air temperature, °C	Measured depth of permafrost (m)
USSR				
Salekhard	67°N, 67°E	50	−7	350
Noril'sk	69°N, 88°E	200	−8	325
Ust'-Port	69°N, 84°E	<50	−11	425
Mirnyy	63°N,114°E	300	−9	550
Tiksi	71°N,129°E	<20	−14	630
Bakhynay	66°N,124°E	50	−12	650
Schalagonzy	66°N,111°E	500	−13	1500
Alaska & Canada				
Resolute	74°N, 95°W	50	−16	390*
Cape Simpson	71°N,155°W	<100	−12	309
Winter Harbour	74°N,111°W	<50	−16	557
Prudhoe Bay	70°N,148°W	<50	−12	*c.* 600

Source: A. I. Yefimov and I. Ye. Dukhin, 1968; R. J. E. Brown, 1972b

*Estimated from extrapolation of temperature gradient in borehole (A. D. Misener, 1955; see p. 462 for further discussion of the data).

were discovered 220 km south of the limit of discontinuous permafrost depicted by R. J. E. Brown in 1960.

Table 2.1 shows some data on permafrost thicknesses in the Arctic. The enormous depths reached by permafrost in northern Yakutskaya (1500 m; P. I. Melnikov, 1966), caused by deep penetration of supercooled brine, are not yet known to be equalled by depths in Canada or Alaska. Drilling for oil at Prudhoe Bay is reported to reveal depths of 600 m of permafrost but greater depths, though not yet proved, are likely to exist. The figure of 557 m at Winter Harbour was obtained from a borehole only 1–2 km inland from the coast of Melville Island, and at greater distances from the sea and at higher elevations, permafrost must be much thicker. R. J. E. Brown (1972a) suggests that in northern Ellesmere Island, for example, at 1000 m elevation where the mean annual air temperature is $-23°C$, the permafrost could be over 1000 m thick. At Braganza Bay, Spitsbergen, ground is frozen in colliery workings to depths of about 320 m, and it extends out beneath the sea for about 100 m from the coast (W. Werenskjold, 1953). Many factors affect the depth of ground freezing; the theory is simple, but precise analysis is extremely complex, as Leffingwell showed as early as 1919. The temperature régime and insolation received at the ground surface (long-term changes must be taken into account), the rate of escape of geothermal heat, and the thermal properties of the ground material (which vary with moisture content) are basic factors. Also to be considered are changes in the thermal properties when ground ice begins to form; the conductivity of dry sand, for instance, can be trebled by formation of interstitial ice. It has long been noted, too, that ground freezes most deeply where the surface is not protected by a cover of vegetation, water, or snow, for all these have a much smaller conductivity than soil or rock. The great depths of frozen ground quoted above are all in the coldest areas of the world, where vegetation is minimal, where summers are short and cool, and where snow does not cover the ground at least until very late in the year. Relatively little is known about the occurrence of permafrost in high mountain regions. J. D. Ives and B. D. Fahey (1971) report on its extent in the Front Range of the Colorado Rockies where, above the tree-line and in areas blown free of snow, it becomes widespread, attaining thicknesses of 60 m.

Temperatures in permafrost are generally lowest near the ground surface, and are influenced by air temperature cycles. Diurnal temperature changes may affect the upper layer to a depth of perhaps 1 m at the most; seasonal temperature changes may be picked up at depths of 15 m or more, though there will be a substantial time lag (up to several months). Temperature fluctuations in the active layer by definition cross the freezing-point regularly; below the permafrost table, however, seasonal tempera-ture fluctuations will take place entirely below freezing-point, and will become smaller with increasing depth until they vanish at the level of 'zero annual amplitude'. At greater depths, temperatures will not fluctuate and will rise steadily with increasing depth until unfrozen ground is reached. An average figure for the geothermal gradient below the level of zero annual amplitude would be 1°C per 30 or 40 m. In the drill hole at Resolute, N.W.T. (F. A. Cook, 1958), seasonal changes amounted to 0·4°C at a depth of 17·6 m (compare the mean annual range of air temperatures: 42°C); the level of zero annual amplitude was 18 m below the surface, but below this the tem-perature continued to fall to a minimum of $-13°C$ at a depth of 30 m. Deeper still, the temperature rose at about 1°C/25·5 m, an unusually steep gradient owing to the thermal effect of the nearby sea, and extrapolation of this gradient suggests a lower limit to the permafrost at about 390 m.

Reversals in the mean thermal gradient may be found, sometimes of sufficient magni-tude to give rise to buried 'taliks'; many factors may be responsible for such a situation but one possible explanation that has always excited interest is that they may reflect long-term climatic changes.

2 The origin of permafrost

There has long been controversy as to the extent to which permafrost is the product of present climatic conditions, and as to its age. Such questions are far from being satisfactorily answered, though many lines of evidence point to an origin for much permafrost in the Pleistocene. As already noted, some temperature profiles show increasing cold at depth and, excluding local causes, a probable explanation is that this represents residual cold that has not yet been dissipated to achieve equilibrium with present climatic conditions. Secondly, although areas of present permafrost and Pleistocene glaciation are not mutually exclusive (R. F. Flint and H. G. Dorsey, 1945), the thickest permafrost is undoubtedly associated with non-glaciated areas. Thirdly, remains of Pleistocene woolly mammoth have been found preserved in permafrost, indicating ground freezing from the time of death to the present. Whether any permafrost survived interglacial periods is unknown. Since the last glacial, the permafrost has adapted slowly to the changing climate so that today much of it is in approximate equilibrium with present temperatures.

In Canada, R. J. E. Brown (1960) claims that there is a very broad relationship between present-day mean annual air temperatures and the zones of permafrost, though there are important discrepancies in some areas. As noted already, permafrost thickness and temperature conditions are affected by several factors, not only air temperature, so that a close relationship is not to be expected; furthermore, data on permafrost occurrence and air temperature records in Canada are still very inadequate. The $-4°C$ isotherm roughly corresponds to the southern limit of discontinuous permafrost in several parts of Canada, though in the Yukon, the $-1°C$ isotherm is a better approximation. South of the $-1°C$ isotherm, permafrost is rare; north of the $-4°C$ isotherm, it is widespread. In the Soviet Union, the boundary between continuous and discontinuous permafrost is correlated with a mean annual ground temperature of $-5°C$, corresponding approximately to a mean annual air temperature of $-8°C$. In eastern Soviet Asia, permafrost extends far south of the $-2°C$ isotherm, and it is thought likely that in this region it is not in equilibrium with the present climate but a relic in part of former cold conditions in the Pleistocene. Permafrost is thicker and more widespread in the USSR, compared with Canada, for several reasons (R. J. E. Brown, 1967). Most important is the fact that far less of the Soviet area was covered by glacial ice in the Pleistocene, whereas nearly all of Canada, excepting the western Yukon, was covered. Thus the low temperatures of the Pleistocene glacial periods were able to penetrate more deeply in the case of the USSR. The latter also experiences a much more 'continental' climate at present. Large areas of Canada were also affected by marine submergence in the post-glacial so that, unlike the USSR, much Canadian low-altitude permafrost has only been exposed to present air temperatures since the end of the last glaciation.

Several workers have tried to correlate permafrost occurrence with climatic variables other than mean annual temperature, using, for example, indices of freezing or thawing. The results of such analyses do not as yet suggest any precise relationship, though J. A. Pihlainen (1962) considered that mean annual air temperature and a thaw index (=degree-days, Fahrenheit, above freezing-point) provided a first approximation to the likely occurrence of permafrost in Canada. Table 2.2 gives data for selected stations in Canada. Further investigations are clearly needed, particularly on the exact occurrence of permafrost at these and other places.

There is good evidence in some places that the present climate is adequate not only to sustain permafrost but actively to form it. Permafrost in new floodplain silts has been frequently observed (for instance, S. Taber, 1943, p. 1504). At Port Nelson, Manitoba, in 1929, a swamp was drained (J. L. Jenness, 1949), its bed being then unfrozen. Within three winters, frost penetrated to a depth of 10 m and there, at a level coinciding with the base of the permafrost in adjacent areas, it stopped, suggesting strongly that the permafrost in this area at least was intimately related to the present climate. Also in the area marginal to Hudson Bay, permafrost now exists in areas that before the post-glacial uplift were beneath the sea and were unlikely then to have been frozen.

On the other hand, the great thickness, location, and intense cold of some southerly permafrost areas in the USSR has led to the view that the permafrost in such areas is essentially 'old' and is now slowly degrading with the amelioration of climate since the last glacial period. Russian investigators suggest that there is an important contrast between the southerly zone of degrading (thawing) permafrost and the far northern zone of active or aggrading permafrost; these two zones are separated by an intermediate belt in which permafrost is being neither actively formed nor actively destroyed by the present climate. The aggrading permafrost may be syngenetic (accompanying

Table 2.2   Comparison of present climate and permafrost occurrence in Canada

Place	Mean annual air temperature °F.	Permafrost occurrence	Thaw index °F. degree-days
Resolute, N.W.T.	3	Continuous (up to 350 m)	540
Churchill, Man.	19	? Continuous (up to 30 m)	1975
Norman Wells, N.W.T.	21	? Continuous (up to 70 m)	3027
Fond du Lac, Sask.	22	None	2789
Great Whale River, P.Q.	23	None	1941
Wabowden, Man.	28	Discontinuous	3373

sedimentation) or epigenetic (forming later than the sedimentation) (I. J. Baranov and V. A. Kudryavtsev, 1963). A. L. Washburn (1973) concludes that the upper part of most continuous permafrost is in approximate balance with present climate, but that most discontinuous or degrading permafrost is either out of balance with present climate or, at best, is in a delicate equilibrium with it. It is in such areas that environmental damage by man may most easily be caused.

It will be apparent that the relationships of permafrost to present-day climate are very complex and far from being fully evaluated. Much permafrost is probably a late Pleistocene relic but has been modified in response to the climate of the last few thousand years or even of the present day. More precise conclusions must await the collection of much more information on the depths and thermal gradients of permafrost.

3 Ground ice

Most ground ice exists in the permafrost of polar and sub-polar lands, and ranges from pore-fillings in sedimentary rocks to masses of clear ice 30 m or more across and the same in thickness. Up to 80 per cent of permafrost by volume may consist of ice. In terms of the proportion of the *weight* of ice to dry soil, the ice content may be over 200 per cent for sections up to 35 m thick in the western Canadian Arctic (J. R. Mackay, 1971) and up to 500 per cent for lesser thicknesses. R. F. Black (1954) distinguishes between rocks which are super-saturated with ice (they contain more ice than pore space), rocks which are saturated (ice volume approximately equal to the pore space: the sediment is cemented by ice, but the ice does not form visible veins or granules), and rocks which are under-saturated. The forms which ground ice takes have been classified by P. A. Shumskiy (1959); a modified version of his classification is given in Table 2.3. Categories 5 and 6 demand no further consideration; needle ice (1*a*) will be mentioned in a later chapter (p. 101). Segregated ice, vein ice, intrusive ice and extrusive ice will be examined in this chapter.

Table 2.3   Classification of ground ice forms

1	Soil ice	*a*	Needle ice (pipkrake)
		b	Segregated ice
		c	Ice filling pore spaces
2	Vein ice	*a*	Single veins
		b	Ice wedges
3	Intrusive ice	*a*	Pingo ice
		b	Sheet ice
4	Extrusive ice		formed subaerially, as in the case of ice formed on river flood-plains (Aufeisen).
5	Sublimation ice		formed in cavities by crystallization from water vapour.
6	Buried ice		buried icebergs, buried glacial ice,etc.

3.1   *Segregated ice*

Segregated ice is a phenomenon of super-saturated rocks. It has been suggested that it should be named 'Taber ice' in honour of its discoverer, S. Taber, who has done so much in the field and in the laboratory to elucidate the mechanisms of ground ice formation. Taber was the first to demonstrate that surface uplift following freezing from the surface downward of certain materials containing water ('frost heaving') was not caused by expansion of water on cooling to form ice, for such expansion of water amounts to no more than 10 per cent whereas heaving needing an expansion of over 100 per cent has often been observed. Taber's experiments (1929, 1930) showed that simple freezing of interstitial water causes practically no uplift, and in any case, not all the water normally freezes except at very low temperatures. His experiments also showed that frost-heaving occurred readily on silts, and less readily on clays, but never on sands. Examination of the frost-heaved materials revealed that uplift was caused by the formation of bands or layers of clear ice ('segregated ice') within the materials; these layers were formed of ice crystals growing within the material in the direction in which heat was being most rapidly conducted away. Ice segregation only occurs in materials now termed 'frost-susceptible', having a suitable grain-size composition (Fig. 2.5). Taber found that quartz dust with a maximum particle size of 0·07 mm was too coarse for ice to segregate; but in materials of grain size 0·01 mm or less, ice segregation could be induced without difficulty, though in clays, segregation might be restricted by the fine pores reducing the rate of water flow to the growing ice crystals. Subsequent workers laid down approximate rules for the determination of frost-susceptibility: for instance, A. Casagrande (1932) suggested 3–10 per cent of the particles should be less than 0·2 mm; but not all frost-heaving samples may conform in practice. This movement of water into the zone in which the crystals are growing is an essential part of the process of ice segregation: the material actually gains in water content during the process, which is not simply a matter of the crystallization of water already in the material. Taber suggested that molecules of water were transferred from adjacent films of water to feed the growing crystals. The water must come largely from ground-water below the layer in which ice is segregating, for the ice and permafrost will effectively prevent any downward percolation from the ground surface; and the water is drawn

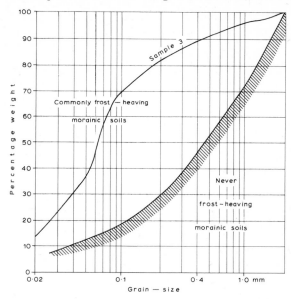

Fig. 2.5
Grain size composition of morainic soils in relation to their susceptibility to frost-heaving (P. J. Williams, *Geogrl J.*, 1957, Royal Geographical Society)

up by tension, not by capillarity since no free water surface exists. Taber estimated the ultimate tensile strength of water by the energy theoretically needed to separate the molecules (when it is converted into a gas) as about 1400 kg/cm². The significance of fine-grained materials in ice segregation is that extremely thin water columns feeding the crystals are more easily maintained under tension than larger ones. In addition, the water remains mobile longer in fine-grained material during freezing, since conversion of interstitial water to ice takes place at a lower temperature in fine-grained than in coarse-grained material. Thus more water can be drawn in to build bigger masses of segregated ice.

Many other factors affect the process of segregation. The shapes of particles are relevant (tabular flat-lying particles are most favourable); very small voids, while assisting the upward lift of water, may delay the freezing process by encouraging supercooling of the water; the rate of cooling is important (very rapid cooling is unfavourable for ice segregation and may even prevent it in some materials); and surface loading may have to be considered, though mainly in artificial conditions (such as civil engineering works), for growing ice crystals can exert a pressure of over 14 kg/cm² (140 t/m²). All these principles were deduced by Taber, and have been later confirmed and elaborated

Plate IV
Ground ice in permafrost, Arctic Canada. Segregated ice is visible in the pit, extending to the right from the blade of the shovel. (C.E.)

by investigators such as G. Beskow (1935) and E. Penner (1960). A recent theoretical study of ice segregation from the viewpoint of soil moisture flow and heat conduction at a frost line is by K. Arakawa (1966), while experiments to simplify the frost susceptibility testing of soils are described by C. W. Kaplar (1971).

The commonest form of segregated ice is that of the ice lens or layer formed parallel with the ground surface. Under favourable conditions, such as slow freezing of water-saturated frost-susceptible materials, ice lenses may grow to great size. Taber (1943) quotes examples 4 m thick from silts in Alaska; J. R. Mackay (1972) reports that recent drilling beneath the arctic coastal plain of northern Canada has encountered widespread ground ice of immense thickness—over 38 m in places. G. B. Cressey (1939) observed ice lenses in shafts at Igarka on the Yenisei river, totalling as much as 7·6 m in thickness out of a section 30 m deep. A. C. Palmer (1967) has discussed why lenses commonly form in series, one above the other, instead of there being continued growth of the first lens. The initial ice lens cannot grow beyond a certain point because of desiccation of the unfrozen soil as water is extracted to feed the growing lens. Palmer shows that, for further ice lenses to form, the thermal diffusivity of the soil must exceed the water diffusion coefficient.

There have in the past been suggestions that the exceptionally massive icy bodies in permafrost of some areas (such as the Canadian Arctic coastal plain) have originated by means other than ice segregation. Hypotheses of burial of glacial ice are now discounted, but the idea of a syngenetic ice-wedge origin has been more strongly held. According to this view, the ground ice formed at the same time as the deposition of sediments by the growth of ice-wedges or vein ice (described in the next section) in those sediments. Mackay (1971) argues against such views that the icy bodies in question are much too large, and that syngenetic ice could not have formed in the marine environment in which the sediments were laid down; he also shows that the ice fabrics supported an origin by ice segregation. The source of the excess water represented by these massive icy beds was from expulsion of ground water during freezing of the sediments, possibly following a lowering of sea level when they were exposed. The term 'epigenetic' is often applied to ground ice of this type that has formed later than the deposition of the sediments.

Massive bodies of ground ice in Quaternary deposits are also described in many parts of the northern USSR. A. I. Popov (1969) reports extensive formations in the Yana-Indigirka lowland which he claims are of syngenetic origin.

The principal morphological effect of ice segregation is differential ground heaving, to be followed, if the ice lenses melt, by differential collapse (see p. 55). The thicknesses of segregated ice vary greatly according to such factors as the amount of available moisture, the rates of cooling (affected by vegetation and snow cover if any), and the grain-size composition. Hummocky relief is typical, consisting of blisters and earth mounds. Differential heaving has serious consequences for any construction works, and it is not confined to permafrost areas—W. H. Ward (1948) recorded 4 cm of frost-heave at Saffron Walden, Essex, in chalk frozen to a depth of only 23 cm in the winter of 1947. Excessive heaving often results from a sudden temperature drop after a spring thaw, for the water content of the soil will be high following melting of segregated ice formed in the winter, and freezing will promote even greater segregation of ice (Taber, 1943, p. 1452). Heaving can affect not only the surface but also individual stones. Suitably shaped stones (e.g., tabular or wedge-shaped downward) may be gripped by an ice lens and slowly lifted, leaving voids beneath them. If on thawing these voids fill with slumped soil, the stones will be unable to drop back into place. A succession of thaw-freeze cycles may eventually lift the stones to the ground surface, a possible basis for the legend common in some cold countries that 'stones grow in the soil'. A. L. Washburn (1973) has over many years studied the process of frost-heaving through experiments in the Mesters Vig district of north-east Greenland (Washburn, 1969). Cone targets and dowels were placed in the ground to various depths and their displacement measured over time. For dry sites in one area (Experimental Site 7) the amount of heaving over one year ranged from zero to about 3 cm, but for wet sites, heaving of up to 8 cm was recorded. The critical factors appeared to be threefold:

1 the moisture content of the soil (wet soils heaved most),
2 the vegetation (most heaving occurred in those places with little or no vegetative cover),
3 the depth of target insertion (deeper targets heaved more).

Washburn develops two hypotheses to explain the heaving of stones to the surface. In the 'frost-pull hypothesis', the ground expands by segregation of ice during freezing and lifts the stones; any spaces beneath them are then filled by slumping or narrowed by frost thrusting. The process has been demonstrated by C. W. Kaplar (1970) in the laboratory using time-lapse photography. The second, 'frost-push' hypothesis states that, because of the greater heat conductivity of stones, ice forms around and beneath them, raising them. On thawing, they do not quite settle back to their original positions, as in the first hypothesis. Washburn suggests that both mechanisms may operate, but that their relative importance may vary according to local site conditions.

Detailed studies of the processes in action raise certain problems and suggest a complex cycle of events. L. W. Price (1972) measured apparent heaving of blocks projecting through turf in an area of drift-mantled slopes, amounts of heaving being about 2·5–5 cm a year. Freshly broken and overturned turf was noted around the bases of the blocks. Paradoxically, most movement was measured in the spring thaw; what was happening was that the surrounding turf and the top soil to a depth of 10–15 cm were moving by gelifluction while the bases of the blocks were still encased in permafrost. Additionally, there was some actual heaving of the blocks, for in the daytime, solar insolation was conducted down through the blocks to thaw a film of the permafrost at their bases in a semi-closed system while, at night, heat loss by radiation caused this moisture to freeze and lift the blocks. After each slight heave, the blocks will never resettle completely in their original positions as already mentioned. Price and others emphasize the importance of the high heat conductivity of stones compared with soil or turf containing air spaces, in promoting rapid temperature changes at the bases of the stones.

3.2   Ice veins and wedges

Vein ice occurs as vertical or near-vertical sheets of ice from less than one to several millimetres thick and sometimes penetrating 10 m or more below the ground surface. Thicker sheets usually taper downward and are known as ice wedges. The largest ice wedges, whose thickness at the top may exceed 10 m, represent another massive form of accumulation of ground ice. Wedges and veins are commonly parts of a polygonal network of ground ice enclosing polygons of frozen ground from 1–30 m in diameter (Chapter 3).

Ice wedges were first described by M. F. Adams (1815) and A. T. von Middendorff (1867) in Siberia. The first correct interpretation of their origin dates back to 1884 (A. von Bunge). A detailed scientific study of ice wedges over a period of some eight years was undertaken by Leffingwell (1915) on the north shore of Alaska, and it is a remarkable tribute to the keenness of his observation that his theory of their origin is still widely accepted today. In the Arctic coastal lowlands and the Yukon river lowlands of Alaska, the ice wedges are formed in 'muck formations', the miners' term for the extensive and thick organic-rich silts. The climate of these regions is characterized by intense winter cold and a large temperature range (for example, Galena: mean January temperature $-24°C$, absolute minimum $-53°C$, absolute maximum $32°C$). The active layer of the muck freezes in winter and contracts on cooling, both in the

Plate V
Frost crack in frozen raised beach sediments, Devon Island, Canada. (C.E.)

active layer and the permafrost beneath. The resulting ground cracking may be audible (Leffingwell, 1915, p. 638). Leffingwell thought that subsequent filling of these contraction cracks with water (later freezing) or ice crystals produced the ice veins, and
further that the cracks provided planes of weakness which would reopen under
stress next winter, allowing deposition of a further film of ice. The maximum width
of cracking in any one winter, he observed, was 8–10 mm. The veins developed

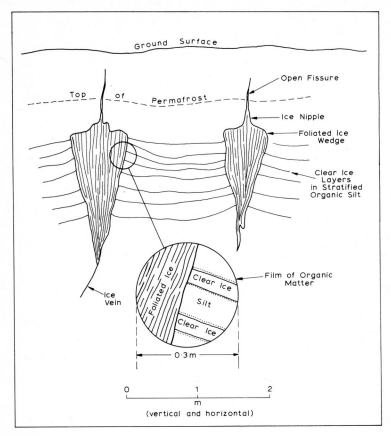

Fig. 2.6
Foliated ice wedges cutting stratified organic silt and clear ice seams, Galena, Alaska (T. L. Péwé,
Biul. Peryglac., 1962)

into ice wedges, the largest of which might take about 1000 years to form on
this basis.

Leffingwell's theory has received support from many later studies. Fig. 2.6 shows
the detailed form of ice wedges studied by T. L. Péwé (1962). The stratification of
the enclosing silts is deformed owing to summer expansion of the ground, as it warms
up and pushes against the ice wedges. The latter show vertical foliation or streaking
caused by fine particles of dirt entering the contraction cracks with the spring meltwater. The detailed mechanics of the process have been examined by A. H. Lachen-

bruch (1962) whose diagram of the evolution of ice wedges is reproduced in Fig. 2.7. As the silts contain very high proportions of ground ice in various forms (Taber, 1943, claimed up to 80 per cent ice content), the rates of contraction and expansion probably differ little from those of clear ice. The coefficient of linear expansion of clear ice is

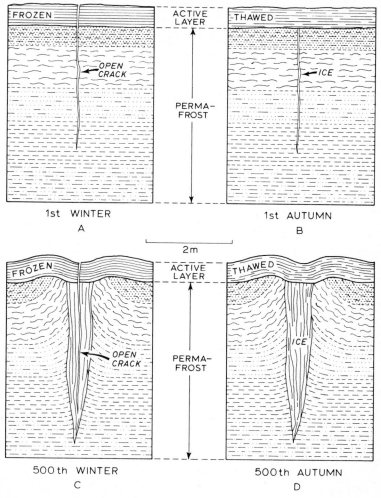

Fig. 2.7
Evolution of an ice wedge according to the contraction crack theory (A. H. Lachenbruch, *Special Paper* 60, 1962, Geological Society of America).

$52 \cdot 7 \times 10^{-6}$ at $0°C$ and $50 \cdot 5 \times 10^{-6}$ at $-30°C$. With a 30°C fall in temperature, cracks about 1·5 mm wide would be expected around a prospective polygon 1 m in diameter. For frozen ground having a coefficient of thermal expansion near to that of ice, a drop in temperature of 4°C may be adequate to cause cracking, but in the case of materials akin to rock, 10°C may be required (R. F. Black, 1963). Observations by D. E. Kerfoot (1972) in the Mackenzie delta corroborate Lachenbruch's theory of thermal cracking.

The pattern of cracking in this area is random except near water bodies, and 80 per cent of the intersections are orthogonal in type, the cracks curving so as to make right-angled intersections.

Frost cracking is occasionally reported from frozen ground in non-arctic regions. A. L. Washburn, D. D. Smith and R. H. Goddard (1963) noticed frost-cracking of a golf-course in New Hampshire following an exceptionally cold December (1958; mean temperature −15°C) with little or no snow cover to protect the ground, which froze to depths of up to 2 m. A network of intersecting cracks appeared, some enclosing polygons 6–30 m across (Chapter 3); the cracks were up to 5 mm wide at the surface. Frost cracking is also known in Iceland where permafrost is not involved (J. D. Friedmann *et al.*, 1971). Water may freeze in the cracks but it melts annually and ice wedges are not formed. Temperature conditions in Iceland are not severe enough even at high levels for ice wedges to form.

J. B. Benedict (1970) has drawn attention to the dangers of concluding that all superficial cracking of the ground following winter cold may result from cooling contraction. The widths of cracks in turf-banked lobes, varying from hair cracks to others 20–30 cm wide, were measured using an invar bar laid across metal pins inserted on either side of a crack. The cracks in this area (Colorado Front Range) were found to widen in autumn and winter and tended to close up in spring and summer. However, the

Plate VI
Pattern of large ice wedges, Foley Island, Baffin Island. (C.A.M.K.)

changes in width did not occur suddenly and showed no correlation with major temperature changes. Hypotheses of gravity movement or desiccation to account for the
cracking were ruled out because of local conditions, and Benedict concluded that, most
likely, the cracks were tension cracks caused by differential frost heaving. Such tension
cracks often become infilled with debris to simulate ice-wedge casts or sand wedges
(see below), but they are quite different and care must be taken to differentiate possible
origins of such superficially similar features before making duductions about past
climates.

A different hypothesis of origin for ice veins and wedges was favoured by Taber
(1943). While admitting that contraction cracks do form (pp. 1447 and 1521), he did
not believe they were responsible for ice-wedge initiation, nor did he accept that they
normally occurred in polygonal patterns. Veins, wedges and lenses were all regarded
as forms of ground ice resulting from segregation, the veins and wedges having grown
downward from ice lenses or layers (p. 1525). The growth of the ice masses themselves
provided the expansive force for cracking the ground. Taber's views on ice-wedge
formation are not now accepted. R. F. Black (1963) makes the interesting comment
that, whereas Leffingwell worked in an area of continuous permafrost, Taber studied
ice wedges in discontinuous permafrost where contraction cracking would be less effective and less evident; the presence of ice wedges in an area of discontinuous permafrost
might even imply its degradation from a former state of continuous permafrost. T. L.
Péwé (1966) argues that ice wedges only form when the mean annual air temperature
is $-6°C$ or colder over a period of many years; the southern limit of present-day active
ice wedges in Alaska roughly corresponds to the mean annual isotherm of $-6°$ to
$-8°C$.

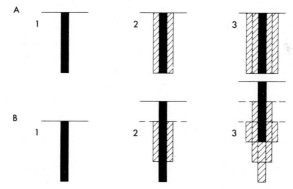

Fig. 2.8
A Epigenetic and **B** syngenetic ice wedges (J. Dylik and G. C. Maarleveld, *Meded. geol. Sticht.*,
1967).

Like some other forms of ground ice, ice wedges may be syngenetic or epigenetic.
Fig. 2.8 compares the theoretical evolution of these types. The former require a longer
period of permafrost while the sediments are accumulating, so they may develop greater
widths with more parallel sides. T. Czudek and J. Demek (1970) refer to examples
of syngenetic ice wedges in Siberia that are 40–50 m wide.

Wedges become extinct when winter stresses at the top of the crack cease to exceed the strength of the wedge ice. This may come about for several reasons. Snow, water, or gelifluction debris may bury the wedge and insulate the ground from large changes of temperature; adjacent wedges may develop to relieve the overall stresses; but if these factors can be eliminated, it is possible that the tops of the wedges buried by later deposits may mark horizons at which important climatic or geomorphological changes took place.

After the ice in a wedge melts, debris will be washed or will slump into the wedge, whose form will be preserved in varying degrees. Fossil ice-wedge casts have been widely recognized in areas that once experienced a severely cold climate. They are often best preserved in silts or clays, whose bedding they intersect cleanly. In sands and gravels, slumping may partially destroy them. G. Johnsson (1959), in a beautifully illustrated paper, gives the following as features useful in the recognition of fossil ice wedges: the filling of sediment must have come from above or the sides, never from below; the sides should converge with depth; stones may stand on end in the wedge in contrast to stones in the surrounding material; and the wedge normally stands more or less vertically. J. Dylik and G. C. Maarleveld (1967) add that the top width is usually more than 10 cm and that the wedges form part of a primary polygonal pattern with diameters mostly greater than 7 m. Care must be taken to distinguish fossil ice wedges from pseudo wedges (for example, chalk pipes, features resulting from glacio-tectonic disturbance, clay squeezed upward by differential loading, or irregular collapses owing to melting of ice lenses). Johnsson (1962) has studied fossil ice wedges in southern Sweden, where the largest measure 6 m in depth and 75 cm wide at the top. R. W. Galloway (1961) lists over 100 wedges in Scotland between the Moray Firth and the Solway, including ice veins up to 5 m deep, and notes how they occur mostly in lowlands outside the area covered by Zone III (post-Allerød) ice. The marked concentration in the north-east is related to this area having been ice-free in the last Würm glaciation (F. M. Synge, 1956). Ice wedges of probable Würm age are known from southernmost Britain; very fine examples were to be seen in clays and sands during the cutting of the M1 motorway near Watford, Hertfordshire. A. V. Morgan (1971) describes fossil ice-wedge casts of epigenetic type in part of the Midlands, where air photography clearly reveals the associated polygonal surface patterns.

Fossil ice-wedge casts must be carefully distinguished from sand wedges. The infilling of the former is an event later than the creation of the ice wedge, whereas the sediment in sand wedges is emplaced during the formation of the crack. Péwé (1959) described how, in the arid region of McMurdo Sound in Antarctica, the contraction cracks are filled with 'clean sand which filters down from above in the spring and summer. Repeated cracking and filling with sand produces a wedge-shaped filling' (p. 550). The sand in the wedges displays vertical lineation, as in the case of ice wedges, and there may be sand-filled veins at the base of the wedge. Some were seen to have formed in stagnant *glacial* ice, showing clearly that they cannot be fossil ice-wedge casts for the glacial ice would itself have disappeared during melting. T. E. Berg (1969) has also reported composite types of wedge filled with a mixture of ice and sand. There thus appears to be a gradation controlled by climate, from the sand wedges of cold arid regions to the ice wedges of cold humid regions, with composite types indicating an intermediate degree of humidity and/or lesser availability of sand.

Sand wedges carry important implications for the interpretation of fossil wedge casts. It should not be assumed that sediment-filled wedges always represent replacements of ice wedges: the alternative is that they are fossil sand wedges formed under cold arid conditions.

4 Some features associated with ground ice

4.1 *Involutions*

Unconsolidated stratified deposits sometimes display contortions of bedding and inter-penetration of one layer by another. These contortions may arise in a variety of ways. Ice shove, mass movements, turbidity currents (*c.f.* P. H. Kuenen's (1953) load-cast structures), differential loading, undermelt (see Embleton and King, 1975, p. 391) and tectonic causes are all possibilities. D. S. McArthur and L. J. Onesti (1970) interpret the contortions found in some Pleistocene sediments near Lansing in Michigan as caused by the foundering of sand laminations into incompetent underlying finer sediments, a response to the quick condition of the sediment immediately following its deposition by running water. However, there still remain many occurrences of contorted structures that are difficult to explain in any of these ways, and processes linked with former frozen ground have been advocated to account for them.

The terms 'involution' and 'injection' have been used for these supposed periglacial features, but not in any consistent way. Many use these terms to refer only to contortions of periglacial origin (as McArthur and Onesti (1970) recommend), others (e.g. J. J. Donner, 1965) use them to cover folded and disturbed structures caused by sub-aqueous sliding and, in some instances, by ice push. The difficulty of restricting the term involution to a purely periglacial connotation is, as Washburn (1973) points out, that so often we cannot prove whether the contortions are the result of periglacial conditions or not.

Periglacial involutions were first described in North America by C. S. Denny (1936) studying contorted sands and gravels in southern Connecticut. R. P. Sharp (1942a) took up the problem when opencast coal mining south-west of Chicago exposed contortions in glacial and fluvioglacial deposits that could not be explained other than in terms of periglacial action. Originally horizontal bedding had been violently deformed where alternations of sand, silt and clay occurred, but not where the sections consisted only of sand. The sand appears to have been intruded in all directions by masses of silt and clay, though downward intrusions predominate. Masses of sand or silt appearing isolated in a section face were in all cases found to be connected in the third dimension (Fig. 2.9); the deformations lacked any linear continuity, each individual fold being quickly replaced by wholly unrelated structures. At their base, the involutions abruptly ceased. Festoons of stones whose long axes follow the contorted bedding often picked out the plugs and tongues of the involutions. Similar descriptions of the phenomenon were given by J. P. Schafer (1949).

A. Jahn (1956) has classified periglacial involutions into three types—fold, pillar and amorphous structures—depending on the degree of disturbance of the deposits. In the third type, the original structure has become so deformed that it is unrecognizable. L.-E. Hamelin and F. A. Cook (1967) differentiate involutions, injections and plications. Injections represent dominantly upward squeezing or migration of fine materials into overlying beds, penetrating and apparently widening cracks in the latter. Plications, or drag folds, are structures induced by down-slope movement, especially gelifluction.

Although a periglacial origin is generally accepted for contorted structures not related to the other causes enumerated above—and particularly in the case of contortions occurring in association with fossil ice wedges (G. Johnsson, 1962)—there is still some uncertainty as to how periglacial involutions were formed. Two principal hypotheses are advocated. One involves the squeezing of a moist plastic layer between rigid

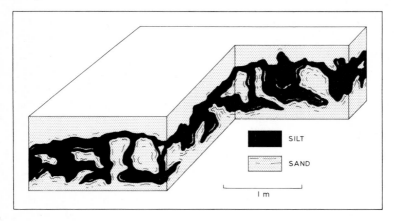

Fig. 2.9
Block diagram to show periglacial involutions (R. P. Sharp, *J. Geol.*, 1942, University of Chicago Press)

permafrost beneath and a newly frozen crust at the ground surface. The permafrost table would mark the downward limit to involutions, which then formed essentially in, and provide a means of determining the presence and thickness of, the active layer. A second possibility, favoured by Sharp, is connected with the process of ice segregation. Frost-susceptible materials, especially silts and clays, expand rapidly in volume on freezing as already explained, with the segregation of ice. Adjacent materials, if still unfrozen, would then be squeezed and intruded by these active centres. Melting of the ice might induce further deformation.

Another hypothesis is that they are related to conditions of hydrodynamic instability in the sediments when the surface layers of the latter thawed and became highly saturated. Adjustments and movements were then initiated by differences in density and viscosity.

Galloway (1961) has studied the distribution of periglacial involutions in Scotland. Twenty-eight sites are recorded, some of which involve contortion of Allerød (Zone II) deposits and the deformations in such cases are therefore likely to be Late-glacial (Zone III). The involutions generally extend down to depths of 0·8–1·8 m, which may be taken as a rough indication of the permafrost table. E. Watson (1965) has compiled information on involutions in west central Wales, south of Towyn, taken to have been ice-free in the last Würm glaciation but subjected to a periglacial climate. Some, he notes, are covered by undisturbed beds and are therefore unlikely to have had any surface expression; but others occur just below present soil level and might have been related to patterned ground at the surface, now destroyed.

Many still automatically associate periglacial involutions with permafrost, but with little or no supporting evidence for this assumption (Péwé, 1969). Probably most cases do reflect the former occurrence of frozen ground, but in others, local freezing and thawing in an active layer may be sufficient to explain them. Their diagnostic value is therefore uncertain.

R. B. G. Williams (1969) maps the occurrence of several periglacial features in southern England, including ice-wedge casts and supposed periglacial involutions, in an attempt to establish the distribution of extensive permafrost (Fig. 2.10). Most of the features on this map south of the Severn–Thames line are involutions, the majority of which seem likely to be periglacial, but Williams points out that if a periglacial origin is not accepted, the area of Britain in which permafrost structures are scarce is greatly extended.

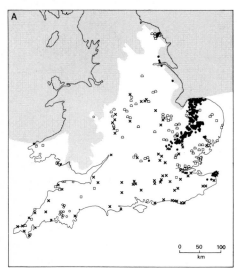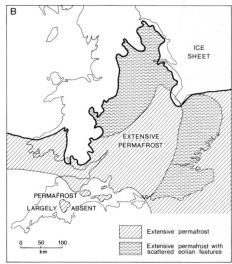

Fig. 2.10
A Occurrence of some periglacial features in southern England. The shaded area was covered by glacial ice in the last glaciation. Involutions are shown by crosses. Infilled circles indicate patterned ground in the Chalklands; open circles denote its occurrence elsewhere. Squares and flat-based semi-circles represent single ice wedges and ice-wedge polygons respectively.
B Distribution of extensive permafrost in the last glaciation deduced from **A**. (R. B. G. Williams in T. L. Péwé (ed.), 1969, *The periglacial environment*).

4.2   *Earth and turf hummocks*

Earth hummocks, sometimes turf-covered, were described by B. Högbom in 1914, and have locally been termed thufurs and buttes gazonnées. They belong to Washburn's

(1956) category of 'non-sorted nets' (see Chapter 3). Their regional occurrence and periglacial relationships have been summarized by C. Troll (1944). R. P. Sharp (1942b) gives a good account of them in the St. Elias Range, Yukon. In this area, they are up to 0·6 m high and 1·5 m in diameter, they are limited to slopes less than 20° (on slopes over 5°, they become elongated across the slope, tending to form small terracettes), and they have earth cores covered with 8–15 cm mats of humus, roots, moss and grass. No sorting of material seems to have occurred and there is no ice in the hummocks. Sharp follows G. Beskow (1935) in suggesting that expanding patches of frozen ground squeeze earth into small knobs. Vegetation in the hollows around and between these knobs tends to be wet and not such a good insulator as the thicker air-filled vegetation on the tops of the mounds which remain unfrozen. Thus the original patchy ground freezing is perpetuated. The influence of other possible factors on the shape and development of the hummocks, such as varying depths of snow cover and snow melt causing swelling of clayey soils, has been discussed for part of the Jura by A. Reffay (1964), who found that the hummocks ('buttes gazonnées') were most commonly situated on slopes where freeze-thaw cycles will be most frequent and effective.

In a present-day arctic region, H. M. Raup (1965) describes turf hummocks on slopes from 1° to 15° in the Mesters Vig district of north-east Greenland. They are composed primarily of mosses which later form the basis for the growth of other plants. All sites of actively growing hummocks are supplied with abundant, gently flowing surface water in summer. Some hummocks have mineral cores, others are built wholly of vegetative and humic matter. They appear to be initiated on micro-elevations such as cobbles, small frost-heaved boulders, or simply on chance irregularities on gelifluction deposits.

Taber (1952) argued that the hummocks were elevated by frost-heaving following segregation of ice lenses beneath. This hypothesis has been strongly refuted by R. S. Sigafoos and D. M. Hopkins (1951, 1954) who agree with Sharp that it is the areas between the hummocks which freeze first, that there is evidence of lateral thrusting or squeezing in the injection of tongues of silt into or beneath the hummocks, and that the fragile nature of some plant roots growing in the hummocks does not accord with a hypothesis of frost-heaving.

A similar phenomenon in parts of western North America is provided by the 'mima mounds', though in many areas these attain somewhat larger dimensions than the hummocks just described. The most famous locality for mima sounds is in the outwash valley plains of the southern Puget Sound region, Washington State. They attain heights of as much as 2 m and may be from 3–20 m in diameter; they are built of a mixture of materials including silt, sand and pebbles. Ever since they were first reported in 1845, there has been speculation and uncertainty as to their origin. The problem became an embarrassment to geological science since all hypotheses prior to about 1940 failed under field tests; some geologists then turned to an origin by gophers, which led to lengthy and unprofitable discussion as to what a gopher might or might not do. It now seems almost certain that the mounds have a periglacial origin, though agreement is by no means reached on the exact processes. A. M. Ritchie (1953), suggests the following sequence of events:

a freezing of the outwash deposits and the formation of ice-wedge polygons;

b thawing of the polygons to a stage where a rounded frozen core still remained surrounded by thawed inter-polygon zones;

c a brief period of erosion by running water, stripping out some of the thawed material.

It may be concluded that there is probably more than one type of earth hummock formed under periglacial conditions, and that a great deal of further work is needed before satisfactory generalizations can be formulated.

4.3 *Palsas*

A rather special form of ground hummock often encountered in present-day arctic and sub-arctic regions is the palsa (Russian: torfyanoy bugor). The term was originally used in northern Finland to describe an ice-cored hummock rising out of a bog (M. Seppälä, 1972a). In periglacial geomorphology, the term is reserved for mounds consisting of peat and ground ice, up to about 10 m high, and up to 10 or 20 m across. They may occur singly but more commonly in considerable numbers in a swamp area, sometimes becoming joined together by winding knobbly ridges (L.-E. Hamelin and A. Cailleux, 1969; S. E. White *et al.*, 1969). The tops of palsas usually consist of a crust of relatively dry peat, which has important insulating properties, preserving ground ice within the mound. Some vegetation may be present. The core may contain peat and/or mineral matter, with ground ice usually in the form of thin layers 2–3 cm thick. According to J. Lundqvist (1962, 1969) the formation of palsas is solely the result of heaving due to ice segregation. In winter, the peat becomes wet, increasing its thermal conductivity; cold penetrates deeply and formation of segregated ice begins. In summer, the drying-out of the surface peat reduces thermal conductivity and the ground ice in the mound is protected so long as this cover of dry peat is not broken. Lundqvist suggests that stages in the growth and decay of palsas, often over a period only of years, may be discerned. Decay sets in when the insulating cover of peat is broken or destroyed. Cracking of the surface may be caused by growth of the palsa itself by thermal contraction in winter, or by desiccation of the peat. Another factor inducing decay could be a rising water level in the surrounding swamp, or summer flooding of the bog with warm water. A dry summer, on the other hand, may cause the vegetation on top of the palsa to die, and wind scour may also help to thin the protective peat cover. After thawing of the palsa, nothing is left—the peat and its stratigraphy resettle. Palsas are typical of areas of sporadic or discontinuous permafrost. The outer limit of the Swedish palsa bogs coincides approximately with the $-2°C$ isotherm (A. Rapp and L. Annersten, 1969).

4.4 *Pingos*

'Pingo' is an Eskimo term for scattered, isolated, dome-shaped hills, and was introduced into geomorphology by A. E. Porsild (1938). Pingos are conspicuous features in the arctic regions of Alaska, Canada, Greenland and Siberia, where there is continuous permafrost. In the USSR they are known as bulgunnyakh. Many have also recently been identified in the sub-arctic boreal forest region of North America, in the zone of discontinuous permafrost. They may be from less than a metre to over 60 m in height, and up to 600 m in diameter. The smaller ones have rounded tops; the larger ones are often ruptured and broken open at the top which has a crater-like appearance as a result. Internally, they usually show outward dipping or deformed beds of stratified silts or sands, though some pingos have developed in bedrock which has been similarly deformed (J. G. Cruickshank and E. A. Colhoun, 1965). In some, an ice core is inferred or visible (R. P. Sharp, 1942c, therefore termed them 'ground-ice mounds'). Unlike palsas, peat plays no part in their formation.

A good example of a pingo is described by F. Müller (1962). Ibyuk Pingo is situated in the Mackenzie delta, N.W.T. It rises 41 m above the surrounding flat marsh, with a basal circumference of 900 m. The summit is in the form of a large crater in which a small lake has collected. It possesses an ice core about 40 m thick, overlain by about 15 m of frozen sediments which have cracked open at the crater. Pingos such as this are, in fact, the most striking relief features of the Mackenzie delta in whose 25,000 km² there occur about 1400, the greatest known concentration of these features. In this area, each occurs in a former lake basin or in a shallow lake at present.

From a series of recent studies, it seems probable that there are at least two main types of pingo, one of which is so well exemplified in the Mackenzie delta region that it is often referred to as the 'Mackenzie type'. In east Greenland, central Alaska, and certain other areas, however, pingos of different origin are termed the east Greenland type. Müller (1959) suggested that the Mackenzie type be termed 'closed-system' pingos, the east Greenland type 'open-system' pingos. It is convenient to discuss their origins separately.

(a) *Mackenzie type*   A comprehensive investigation of pingos in the Mackenzie delta has been undertaken by J. R. Mackay (1962). They are not simply the result of ice-lens heaving since they are formed in material which is 95 per cent fine to medium sand (0·1–0·5 mm), with very little silt or clay, and therefore not significantly frost-susceptible. On the other hand, the sand allows rapid movement of pore-water. Fig. 2.11 illustrates Mackay's theory of their formation. The initial situation comprises a relatively large deep ice-covered lake surrounded by permafrost, while beneath the central position of the lake there is unfrozen ground. As the lake is slowly infilled with sediment, there comes a point when the lake ice becomes frozen to the bottom, and the bottom sediments in turn freeze, so that a layer of ice and permafrost extends over the site of the lake. Thus there is created a 'closed-system' in the unfrozen ground beneath this new cap of permafrost from which water cannot escape. Inward growth of permafrost around the unfrozen core increases water pressure in the latter, for pore-water is expelled from the unfrozen sands by the advancing frost plane. To relieve the pressure, the surface layers are bulged up. Eventually, all the water in the closed system is converted to ice and the excess water forms a core of clear ice under the bulge.

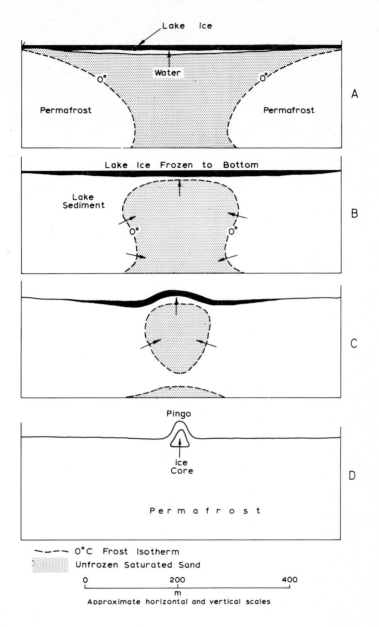

Fig. 2.11
Pingo formation in the Mackenzie Delta area, according to J. R. Mackay (*Geogr. Bull., Ottawa*, 1962, by permission of the Queen's Printer, Ottawa). Depth of lake ice and water exaggerated;
A Broad shallow lake over frost-free ground;
B Accumulation of sediment in lake causes lake ice to freeze to bottom in winter. Edge of permafrost advancing, causing expulsion of pore-water in the saturated sands and giving upward pressure beneath lake centre;
C Further advance of permafrost; updoming of lake ice and sediment;
D Permafrost continuous; up-doming has ceased, and ice-cored pingo remains

The rate of pingo formation appears to be variable. In the early stages, growth may be surprisingly rapid considering the fact that the ice cores require large volumes of unfrozen sediments from which the water is gradually expelled. Mackay (1973) measured the growth of some pingos in north-west Canada by precise levelling, finding a maximum rate of 1·5 m/yr for the first one or two years, after which the rate rapidly decreases. Several pingos in the Mackenzie delta are known to have formed since 1950. Large pingos may continue to grow slowly for over 1000 years. Radiocarbon dates on two pingos suggest average growth rates of 0·3–0·5 m per 1000 years. They are thought to be stable features of permafrost areas so long as the surface remains intact, but the up-bulging helps to cause cracking of the surface, allowing exposure of the ice core and its eventual disappearance by melting when air temperatures permit. The centre of the pingo then collapses to form a crater.

A completely different view of the mode of formation of the Mackenzie pingos has been put forward by R. C. Bostrom (1967). He regards them as being of the east Greenland type, and it is more appropriate to consider his views, together with Mackay's refutation of them, in the next section.

(b) *East Greenland type*   Leffingwell in 1919 suggested that isolated mounds on the gently sloping coastal plain of north Alaska were formed by hydraulic forces. Progressive downward freezing of water-saturated sediments might interfere with normal ground-water flow and cause sub-surface pressure to build up. In extreme cases, the ground would be bulged up and sometimes even ruptured, giving an outburst of water and debris. The general requirements for 'open-system' pingos are that there should be permafrost, and an aquifer beneath the permafrost by which water flows to the pingo site.

Bostrom (1967), as noted above, has recently reinterpreted the Mackenzie pingos as being of this hydraulic type. He adduces evidence to show that the Mackenzie delta is an area of contemporary subsidence. Sediment is being added and frozen on to the permafrost layer while, at the base of the latter, sediment is thawing on being carried by subsidence through the 0°C isothermal surface. Thus a rigid frozen layer, possibly 90 m thick, rests on a newly thawed and water-saturated zone below; the resulting hydraulic pressure will be 11·7 kg/cm^2 if the mean density of the frozen layer is taken to be 1·3. In places where the frozen layer is cracked (for instance, as a result of flexure) or otherwise interrupted, water (density 1·0) will rise from the thawed zone by artesian pressure to a theoretical height of 27 m above the ground surface. The figure of 27 m is approximately in accordance with the mean height attained by the Mackenzie pingos. As the water moves upward through the permafrost, its relative warmth maintains and enlarges a thawed passage and forms a pond at the ground surface. When the pond shallows sufficiently by silting, it can become frozen to its bottom in winter, and continued upward movement of artesian water will dome up the ice to form a pingo. The ponds or lakes associated with the Mackenzie pingos are thus explained not as initial features (Mackay, 1962) but as consequent on the upward movement of water under hydraulic pressure.

area in which most of the pingos in this part of Canada lie is east of the Mackenzie delta, is never flooded by the Mackenzie river and is not undergoing subsidence and sedimentation. These pingos could not have formed contemporaneously with sedimentation, for the deposits are Pleistocene and the pingos would have to have survived for tens of thousands of years. In the Mackenzie delta proper, pingos are relatively few and concentrated in the seaward zone of newly forming and thin permafrost. Secondly, the permafrost is interrupted by many gaps ('like a perforated sieve') and long-distance sub-surface flows of expelled pore-water are not likely. Thirdly, it seems unlikely that expelled and rising pore-water could transfer enough heat to keep thawed chimneys open in the presence of surface heat losses. Fourthly, recent boreholes show that the permafrost in the main pingo area is over three times as thick as Bostrom assumes.

G. W. Holmes, D. M. Hopkins and H. L. Foster (1963) believe that many pingos in central Alaska belong to the east Greenland category. Most lie on gentle forested slopes in unglaciated silt-filled valleys; large ice lenses within them are sometimes exposed. Their ages are thought to vary from some decades up to about 7500 years. Further examples in north-east Greenland are given by Cruickshank and Colhoun (1965), while in central Yukon, O. L. Hughes (1969) has identified numerous open-system pingos, most abundant in non-glaciated areas or areas unaffected by Wisconsin ice.

In a study of pingos on Prince Patrick Island, N.W.T., A. Pissart (1967) differentiates two distinct types of pingo, and claims that they belong neither to the Mackenzie nor to the east Greenland type. One group occurs on high ground far from any lakes and seems to be related to groundwater moving up faults in the bedrock. Another group of very elongated pingos, located on coastal or valley-floor sediments, is of uncertain origin, but possibly related to thawing or re-forming of permafrost as post-glacial sea-level changes caused the sea to invade or retreat from these low-lying areas.

H. M. French (1971) has described ice-cored mounds 'or micro-pingos' that occur within the low central areas of ice-wedge polygons on poorly drained meadow Tundra soils. Ice segregation may be brought about by contraction of the water-saturated layer under cryostatic pressure. The upstanding ice wedges bounding the polygons form the boundary of a closed system. They are similar in some respects to the Mackenzie delta type pingos. The polygons have four to five sides and diameters up to 60 m. The mounds in the polygons are 2–5 m in diameter and none in the area studied exceeded 50 cm in height above standing water. All had cracks in the tundra mat and had clear solid ice inside, some showed signs of collapse. The permafrost was 25–30 cm below the standing water, but the vegetation on the mound core rested straight on ice. The ice wedges rose 25 cm above the water surface. The mounds superficially resemble palsas, but occur on permafrost and have solid ice cores. They exist in all stages of cyclic development from embryonic to collapse forms, so that the cycle may be relatively short.

It seems clear that there may be greater variety in pingo forms and origins than has hitherto been revealed, and that while detailed studies have been made of the so-called Mackenzie type, the open-system and other pingos have received less attention.

Submarine pingos have recently been described from the Beaufort Sea (J. M. Shearer *et al.*, 1971); they may be forming today, for sea-water temperatures at the sea bed are as low as $-1\cdot8°C$, so that freshwater in the submarine sediments could freeze.

(*c*) *Fossil pingos* have been recorded by many, though some claims are dubious and there is great risk of confusion with other features. When the pingo ice core melts away, a closed depression, sometimes surrounded by a low rampart, may be left. The ramparts may be useful in enabling one to distinguish fossil pingo depressions from other hollows such as kettles, solution-subsidence features, thermokarst depressions (p. 59) or man-made depressions.

Cailleux (1956) suggested that the *mares* of the Paris Basin might be fossil pingo depressions. On the high plateaux of southern and eastern Belgium, hundreds of fossil pingos are said to be present. Deposits in them date back to the Late-glacial—older ones have presumably been obliterated by gelifluction or loess (W. Mullenders and F. Gullentops, 1969). Their diameters are up to 120 m; on slopes, the ramparts become elongated and interrupted on the up-slope side (A. Pissart, 1963). G. F. Mitchell (1971) recognizes similar forms in the south of Ireland, with rims raised about $1\cdot5$ m, and E. Watson (1972) has mapped their distribution in Cardiganshire, Wales. The basins contain grey clay-silt. All lie beyond the limit of the last glaciation according to Mitchell and Watson. Although modern pingos are not so far known in Scandinavia, fossil forms, often containing small lakes, are described by H. Svensson (1969) in northern Norway and by Seppälä (1972b) in Finnish Lapland. G. Wiegand (1965) claims examples in parts of south and central Germany.

With such a widespread apparent distribution and existence supported by so many workers, the reality of fossil pingos is hardly in doubt, but their precise differentiation from other morphologically similar features creates problems. The most reliable test is in studies of internal structure, such as J. Dylik (1965) carried out in the Łodz area of Poland.

4.5 *Extrusive ice*

A. T. von Middendorff in 1859 (Leffingwell, 1919) first named and explained interesting forms of ground ice that he called 'Aufeisen'. The ones that he discussed were associated with river floodplains. In winter, river flow is impeded by freezing in the form of anchor ice at the bottom and frazil ice at the surface of the river, and shallow stretches may freeze solidly. Water still flowing upstream floods the area as the flow becomes increasingly impeded. Freezing then forms candle ice which consists of long vertical ice crystals. Alternate flooding and freezing, continuing all winter, may build up layers of candle ice 4 m thick and covering up to 8 km². Domes and ridges form and these may crack as they are forced up by water pressure from below.

Aufeisen have been found in west Baffin Island in association with large kettle-holes. Where kettle-holes exceed $0\cdot7$ km in diameter and are deep, the water cannot freeze from top to bottom. The kettle-holes associated with the Aufeisen have no visible outlets and the water must escape through the coarse bouldery beds to feed springs which occur near the Aufeisen. The spring water escapes at intervals during the winter when the pressure exceeds the strength of the ice forming at its margin. The water then floods out over the Aufeisen. Some layers of candle ice consist of candles 42 cm long. One Aufeis was fed by a small kettle-hole with no visible inflow but a strong outflow of water. The small kettle-hole was fed in turn by a much larger one a short distance away and at a higher level. This Aufeis occupied a small gully and was about 1 km long with a gradient of 3·5 degrees; it was at least 3 m thick. It consisted of horizontal layers of candle ice interbedded with compressed snow which must have fallen between outbursts. The most distinctive feature of Aufeisen is the candle ice of which they mainly consist, and the essential requirement for their formation is the periodic availability of water throughout the winter.

4.6   Thermokarst

In karst areas, a distinctive set of landforms is related to processes of solution acting upon limestone or dolomite bedrock. 'Thermokarst' is a term that also applies to a peculiar group of landforms and the action of a particular process in producing them, but in this case it is not a chemical but a physical process. However, there are some striking points of resemblance between karst and thermokarst landscapes, as H. Svensson (1970) has described. The term 'thermokarst' was introduced by a Russian, M. M. Yermolayev, in 1932, to describe the features resulting from the melting of ground ice in permafrost regions. When the permafrost thaws, a highly mobile liquid sludge results, and the thawed ground, especially if the permafrost is supersaturated (see p. 33) has a much smaller volume, perhaps only a small fraction, compared with the former permafrost. The principal effect is therefore to create collapse features—surface pits and basins (largest where the content of mineral matter in the original permafrost was least), funnel-shaped sinks, dry valleys, or ravines; before collapse occurs, caves may even be found where ice has begun to melt away. For its best development, thermokarst needs numerous and thick ground-ice masses in the upper parts of the permafrost. General discussions of the phenomena are given by A. I. Popov (1956) and J. Dylik (1964). S. P. Kachurin (1962) recognizes three zones of thermokarst in the USSR, in which the southernmost represents areas where degradation of the permafrost is today active and where thermokarst is extensively developed, and the northernmost represents a region where there is no degradation of permafrost but where the presence of abundant surface water at times (for example, spring floods) leads to locally intense thawing of ground ice. He and other writers also note how human activities such as ploughing, ditch digging, or building operations in permafrost may greatly accelerate thermokarst development.

A detailed discussion of the range of thermokarst forms is given by Czudek and Demek (1970, 1971, 1973). The following categories are distinguished:

(a) *Thermo-abrasion forms*   These are associated with the shorelines of seas and lakes where the combined mechanical and thermal effect of wave-action on the permafrost results in notching and rapid undercutting of the land. Fig. 2.12A gives some examples on the Soviet Arctic coast, and Table 2.4 shows stages in the gradual disappearance of an entire island.

Table 2.4 Thermo-abrasion and the island of Semenowski ostrow, Arctic Ocean

Year	Length (km)	Width (km)
1823	14·8	4·6
1912	4·6	0·9
1945	1·6	0·2
1956	island disappeared	

(*b*) *Thermo-erosion forms* The term is restricted to the action of flowing river water which, because of its relative warmth, undercuts frozen banks at the waterline and causes bank recession. The term 'thermo-erosion niche' was proposed by A. I. Gusev (1952) for the undercut, and D. Gill (1972) has illustrated the phenomenon in the Mackenzie River delta. Czudek and Demek (1970) describe how ice wedges thaw out from the river bank inland, and the centres of ice-wedge polygons gradually become isolated by this process to form conical hills known as baydjarakhs in Yakutia (compare the 'mima mounds' discussed on p. 48). The baydjarakhs are then themselves slowly destroyed.

Plate VII
Thermo-erosion and slumping of the permafrost bank of the Mecham River, Cornwallis Island, Canada. (C.E.)

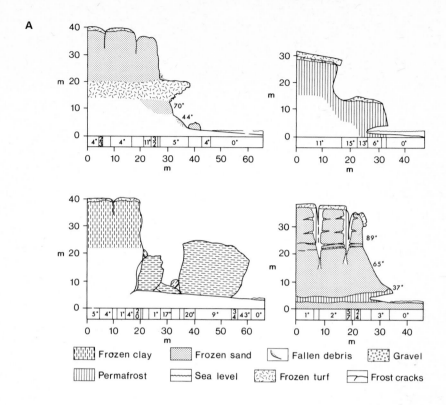

A

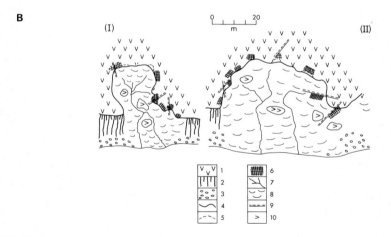

B

Fig. 2.12
A Effects of thermo-abrasion on Soviet Arctic coast profiles (T. Czudek and J. Demek, *Rozpravy Československé Akademie Věd*, 1973);
B Development of thermo-cirques. 1, Taiga-covered terrace surface; 2, Steep terrace edge; 3, River gravels; 4, Edge of thermo-cirque; 5, Incipient thaw zones; 6, Exposed ice wedges; 7, Rills; 8, Bulges of softened clay; 9, Orientation of visible ice wedge; 10, Slipped blocks of frozen clay

(*c*) *Thermo-suffosion forms* result from the thawing of ground ice and the accompanying mechanical washing-out of fines in the groundwater draining away, leading to collapse features at the surface.

(*d*) *Thermo-planation forms* may develop in two ways. First, lateral degradation from river banks or lake shores, as ice wedges are melted out as described under (*b*), may reach a stage when the baydjarakhs are destroyed and amphitheatre-shaped hollows (termed thermo-cirques) are created (Fig. 2.12B). Pools of water may collect in these. Secondly, downward degradation of a generally flat permafrost surface may take place in a similar way by the selective melting of ice wedges. Fig. 2.13 shows the diagrammatic evolution, leading to the stage when alas lakes are formed. The role of these water

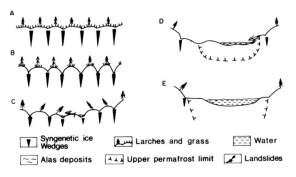

Fig. 2.13
Degradation of permafrost and the evolution of alases (T. Czudek and J. Demek, *Quaternary Research*, 1970)

pools is very important since the water stores the heat needed for the thaw process. Once the alas lake is deep enough, it ceases to freeze to the bottom in winter and a talik forms in the permafrost. Some alas lakes in Yakutia are up to 40 m deep and 15 km across. They eventually fill up with sediment or are drained to other, lower, alases; then permafrost aggradation begins again. At this stage, pingos and palsas may form on the floor. The thermokarst valleys that link alases and drain them, form complex networks with unexpected blind ends and stepped profiles.

In Alaska, some thermokarst forms were recognized in the late 1940s by R. E. Wallace (1948) and D. M. Hopkins (1949). Both discuss the development of 'thaw lakes' and collapse basins. The features can be closely matched with the various forms categorized by Czudek and Demek: what Hopkins and Wallace term thaw lakes are analogous to the Siberian alases. They describe how an insulating cover of vegetation is broken by frost-heave; patchy thawing of ice-rich sediments will then give rise to scattered hollows containing pools of water. Once formed, such pools steadily enlarge themselves (Hopkins notes 5-metre thermo-erosional undercuts). In the early stages, lakes are small and scattered, sometimes even occurring on flat hill-tops, and drainage is extremely haphazard. Subsequently, the lakes begin to coalesce as they grow in size, and integration of drainage commences. Wallace finds evidence for a still later stage when through-flowing streams begin to build levees and to separate thaw lakes into segments. The lakes also enlarge along ice wedges (C. E. Carson and K. M. Hussey, 1959), giving thaw gullies extending a hundred metres or more from the lake shore. Sometimes lakes may be drained underground giving 'thaw sinks'. Rates of growth of thaw lakes from 5 to 20 cm a year were determined by Wallace by noting the progressive changes in the trees affected by bank collapse—tilted, bent, and S-shaped trees occur. A lake 100 m in diameter might need only 500–1500 years to form.

R. F. Black (1969) has summarized North American data on thaw lakes. Their size and rate of formation depend closely on the ice content of the frozen ground. Permafrost in northern Alaska, for instance, has an ice content in gravelly material of 16–18 per cent by volume when supersaturated; in sand the proportion may vary from 9–40 per cent; in silt, 13–65 per cent; and in clayey silt and silty peat from 75–91 per cent. Ice wedges are additional to these figures: some wedges attain widths of 5 m and locally as much as 50 per cent of the upper 6 m of the permafrost may be taken up with wedge ice. Thus ground-ice content, including both segregated ice and wedge ice, may theoretically approach 100 per cent in the upper layers, although such a condition has not been recorded. Thaw depressions or lakes form rapidly and attain larger dimensions when there is a relative paucity of mineral matter in the active layer. Some in eastern Alaska extend by bank collapse at up to 19 cm/year.

Fossil thermokarst basins have been described from some Pleistocene periglacial areas. P. E. Wolfe (1953) mapped about 2500 closed depressions up to 6 m deep and 2·5 km² in size on the coastal plain of New Jersey. The hollows occur on various rock types from clays to gravels. All hypotheses of origin, other than one of thaw-lake development when the ground was perennially frozen, were found unsatisfactory. Similar forms have been located in the Paris Basin by Pissart (1958) . The circular depressions here are normally found on impermeable rocks, and neither karstic nor weathering processes can adequately explain their occurrence though an alternative view that some may be fossil pingo depressions (A. Cailleux, 1956) has already been noted. Pissart suggests that they are thermokarst basins, developed when sub-surface ice lenses melted away accompanied by slow surface subsidence.

It is useful at this point to deal briefly with enclosed depressions of possible periglacial origin as a whole. Two main periglacial hypotheses have been presented: that the depressions may be fossil pingos, and that they may be fossil thermokarst basins (thaw lakes). There are, as has already been stressed, innumerable non-periglacial hypotheses of origin, ranging from the man-made (such as the marl pits of eastern England) to karstic (such as the gentle dimples of solution-subsidence on the chalk of north-west Europe) and fluvioglacial (kettle-holes). Determining which hypothesis is most plausible is far from easy even if adequate sections can be dug. The presence of raised rims has been taken by some to be diagnostic of a fossil pingo origin, but any ground-ice lens must displace material laterally around it, as B. W. Sparks et al. (1972) observe, so that a thermokarst hypothesis is not ruled out. In the case of the Breckland meres of eastern England, Sparks et al. (1972) argue against a fossil pingo origin on the grounds that the hollows do not occur in Mackenzie-type lake sites, nor in artesian sites of the east Greenland type. Their infillings suggest that the Breckland depressions formed in Late-glacial cold periods and a thermokarst origin is advocated.

5 Conclusion

Permafrost is defined as a thickness of bedrock or superficial material in which temperatures are below freezing over a period of years. Zones of continuous permafrost are mainly related to areas of present-day severely cold climate, though some parts may be relict from the Pleistocene. Maximum known depths of permafrost approach 1500 m in the northern USSR and 600 m in northern Alaska, regions of intense winter cold, short cool summers, minimal vegetation cover, and small snowfall. Above the permafrost lies the active layer which experiences summer or short-period thawing. Buried thawed layers (taliks) within permafrost may reflect long-term climatic changes, or merely the influence of local factors such as pockets of highly mineralized ground water.

Ground ice in permafrost ranges from pore-fillings in sedimentary rocks to large ice lenses and wedges. Segregated ice has been studied in detail by S. Taber and others, and it has been shown that it forms in frost-susceptible materials, having a suitable grain-size composition, when ground water is drawn in to feed growing ice crystals at a frost line. Given slow freezing of water-saturated frost-susceptible materials, lenses or layers of segregated ice formed parallel to the surface may reach several metres in thickness, resulting in localized heaving of the ground surface. Veins and ice wedges form in semi-vertical cracks which, as E. K. Leffingwell postulated sixty years ago, result from thermal contraction of frozen ground during cooling. They probably require mean annual air temperatures below $-6°C$. After the ice in a wedge melts (for a variety of possible reasons), debris may slump in and preserve the structure of the wedge in fossil form. In areas of cold arid climate, sand wedges represent an analogous feature. Other features associated with ground ice include involutions, probably caused by processes of ice segregation, and pingos, which are dome-shaped mounds up to 60 m high produced by freezing of ground water trapped beneath permafrost or lake ice. Pingos probably require a less severe climatic régime than ice wedges, for in Alaska they are found up to the mean annual isotherms of $-1°$ or $-2°C$. Another form of ground ice is classified as extrusive ice and includes Aufeisen formed by periodic escape of spring or river water in winter.

The decay of ground ice, resulting mainly from climatic amelioration, gives rise to thermokarst features, such as pits and basins caused by collapse of layers once supported by ground ice. Thaw lakes and alas basins are another manifestation of decaying ground ice.

6 References

ADAMS, M. F. (1815), see letter by A. CAILLEUX, *J. Glaciol.* **10** (1971), 411

ANDERSON, G. S. and HUSSEY, K. M. (1963), 'Preliminary investigations of thermokarst development on the North Slope, Alaska', *Proc. Iowa Acad. Sci.* **70**, 306–20

ANDREWS, J. T. (1961), 'Permafrost in southern Labrador-Ungava', *Can. Geogr.* **5**, 34–5

ARAKAWA, K. (1966), 'Theoretical studies of ice segregation in soil', *J. Glaciol.* **6**, 255–60

BARANOV, I. J. and KUDRYAVTSEV, V. A. (1963), 'Permafrost in Eurasia', Proc., Permafrost Int. Conf., Indiana (Lafayette, 1963). *Nat. Acad. Sci.—Nat. Res. Council Publ.* **1287**, 98–102

BENEDICT, J. B. (1970), 'Frost cracking in the Colorado Front Range', *Geogr. Annlr* **52**A, 87–93

BERG, T. E. (1969), 'Fossil sand wedges at Edmonton, Alberta, Canada', *Biul. Peryglac.* **19**, 325–33

BESKOW, G. (1935), 'Tjälbildningen och tjällyftningen med särskild hänsyn till vägar och järnvägar', *Sver. geol. Unders. Afh.*, Ser. C, **375** (English translation by J. O. OSTERBERG (1947), *NWest. Univ. tech. Inst.*, 145 pp.)

BLACK, R. F. (1954), 'Permafrost—a review', *Bull. geol. Soc. Am.* **65**, 839–55

(1963), 'Les coins de glace et le gel permanent dans le Nord de l'Alaska', *Annls Géogr.* **72**, 257–71

(1964), 'Periglacial studies in the United States, 1959–63', *Biul. Peryglac.* **14**, 4–29

(1969), 'Thaw depressions and thaw lakes. A review', *Biul. Peryglac.* **19**, 131–50

BOSTROM, R. C. (1967), 'Water expulsion and pingo formation in a region affected by subsidence', *J. Glaciol.* **6**, 568–72

BROWN, R. J. E. (1960), 'The distribution of permafrost and its relation to air temperature in Canada and the USSR', *Arctic* **13**, 163–77

(1967), 'Comparison of permafrost conditions in Canada and the U.S.S.R.', *Polar Rec.* **13**, 741–51

(1972a), 'Permafrost in the Canadian Arctic archipelago', *Z. Geomorph.*, *Suppl.* **13**, 102–30

(1972b), *Permafrost in Canada* (Toronto), 234 pp.

BRYAN, K. (1946), 'Cryopedology—the study of frozen ground and intensive frost-action with suggestions on nomenclature', *Am. J. Sci.* **244**, 622–42

BUNGE, A. VON (1884), see letter by A. CAILLEUX, *J. Glaciol.* **10** (1971), 411

CAILLEUX, A. (1956), 'Mares, Mardelles et pingos', *C. R. Acad. Sci.* **242**, 1912–14

CARSON, C. E. and HUSSEY, K. M. (1959), 'The multiple working hypothesis as applied to Alaska's oriented lakes', *Proc. Iowa Acad. Sci.* **66**, 334–49

CASAGRANDE, A. (1932) in BENKELMAN, A. C. and OLMSTEAD, E. R. 'A new theory of frost heaving', *Natn. Res. Coun. Highw. Res. Bd, Proc. 11th Ann. Mtg.* **1**, 168–72

COOK, F. A. (1958), 'Temperatures in permafrost at Resolute, N.W.T.', *Geogr. Bull.* **12**, 5–18

CRESSEY, G. B. (1939), 'Frozen ground in Siberia', *J. Geol.* **47**, 472–88

CRUICKSHANK, J. G. and COLHOUN, E. A. (1965), 'Observations on pingos and other landforms in Schuchertdal, north-east Greenland', *Geogr. Annlr* **47**, 224–36

CZUDEK, T. and DEMEK, J. (1970), 'Thermokarst in Siberia and its influence on the development of lowland relief', *Quat. Res.* **1**, 103–20

(1971), 'Der Thermokarst im Ostteil des Mitteljakutischen Tieflandes', *Scripta Fac. Sci. Nat. Ujep Brunensis, Geographia* **1**, 1–19

(1973), 'Die Reliefentwicklung während der Dauerfrostbodendegradation', *Rozpr. čsl. Akad. Věd.* **83**, 69 pp.

DENNY, C. S. (1936), 'Periglacial phenomena in southern Connecticut', *Am. J. Sci.* **32**, 322–42

DONNER, J. J. (1965), 'The Quaternary of Finland' in *The Quaternary* (ed. K. RANKAMA), **1**, 225

DYLIK, J. (1964), 'Le thermokarst, phénomène négligé dans les études du Pleistocène', *Annls Géogr.* **73**, 513–23
(1965), 'L'étude de la dynamique d'évolution des dépressions fermées à Józefów aux environs de Łódz', *Revue Géomorph. dyn.* **15**, 158–71

DYLIK, J. and MAARLEVELD, G. C. (1967), 'Frost cracks, frost fissures and related polygons: a summary of the literature of the past decade', *Meded. geol. Sticht.* **18**, 7–21

EMBLETON, C. and KING, C. A. M. (1975), *Glacial geomorphology*

FERRIANS, O. J. KACHADOORIAN, R. and GREENE, G. W. (1969), 'Permafrost and related engineering problems in Alaska', *U.S. geol. Surv. Prof. Pap.* **678**, 37 pp.

FLINT, R. F. and DORSEY, H. G. (1945), 'Glaciation of Siberia', *Bull. geol. Soc. Am.* **56**, 89–106

FRENCH, H. M. (1971), 'Ice-cored mounds and patterned ground, South Banks Island, West Canadian Arctic', *Geogr. Annlr* **53**A, 115–45

FRIEDMANN, J. D. *et al.* (1971), 'Observations on Icelandic polygon surfaces and palsa areas', *Geogr. Annlr* **53**A, 115–45

GALLOWAY, R. W. (1961), 'Ice wedges and involutions in Scotland', *Biul. Peryglac.* **10**, 169–93

GERASIMOV, I. P. (1961), 'The recent nature of the Siberian pole of cold', *J. Glaciol.* **3**, 1089–96

GILL, D. (1972), 'Modification of levee morphology by erosion in the Mackenzie River delta, Northwest Territories, Canada', *Inst. Br. Geogr. Spec. Publ.* **5**, 123–38

GRIPP, K. (1927), 'Beiträge zur Geologie von Spitsbergen', *Abh. Geb. Naturw., Hamburg* **21**, 1–38

GUSEV, A. I. (1952), 'On the methods of surveying the banks at the mouths of rivers of the polar basin', *Trans. Inst. Geol. Arctic* (Leningrad), **107**, 127–32

HAMELIN, L.-E. and CAILLEUX, A. (1969), 'Les palses dans le bassin de la grande-rivière de la Baleine', *Rev. Géogr. Montréal* **23**, 329–37

HAMELIN, L.-E. and COOK, F. A. (1967), *Illustrated glossary of periglacial phenomena* (Quebec)

HÖGBOM, B. (1914), 'Über die geologische Bedeutung des Frostes', *Bull. geol. Instn Univ. Upsala* **12**, 257–389

HOLMES, G. W., HOPKINS, D. M. and FOSTER, H. L. (1963), 'Pingos in central Alaska', *Geol. Soc. Am. Spec. Pap.* **73**, 173

HOPKINS, D. M. (1949), 'Thaw lakes and thaw sinks in the Imuruk Lake area, Seward Peninsula, Alaska', *J. Geol.* **57**, 119–31

HOPKINS, D. M. and SIGAFOOS, R. S. (1954), 'Role of frost thrusting in the formation of tussocks', *Am. J. Sci.* **252**, 55–9

HUGHES, O. L. (1969), 'Distribution of open-system pingos in Central Yukon Territory with respect to glacial limits', *Geol. Surv. Can. Pap.* **69–34**, 8 pp.

IVES, J. D. (1960), 'Permafrost in central Labrador-Ungava', *J. Glaciol.* **3**, 789–90

IVES, J. D. and FAHEY, B. D. (1971), 'Permafrost occurrence in the Front Range, Colorado Rocky Mountains, U.S.A.', *J. Glaciol.* **10**, 105–11

JAHN, A. (1956), 'Some periglacial problems in Poland', *Biul. Peryglac.* **4**, 169–83
 (1958), 'Periglacial microrelief in the Tatras and on the Babia Góra', *Biul. Peryglac.* **6**, 227–49

JENNESS, J. L. (1949), 'Permafrost in Canada', *Arctic* **2**, 13–27

JOHNSSON, G. (1959), 'True and false ice-wedges in southern Sweden', *Geogr. Annlr* **41**, 15–33
 (1962), 'Periglacial phenomena in southern Sweden', *Geogr. Annlr* **44**, 378–404

KACHURIN, S. P. (1962), 'Thermokarst within the territory of the U.S.S.R.', *Biul. Peryglac.* **11**, 49–55

KAPLAR, C. W. (1970), 'Phenomenon and mechanism of frost-heaving' in *Frost action: bearing, thrust, stabilization and compaction* (Highway Research Board Record **304**), 1–13

KERFOOT, D. E. (1972), 'Thermal contraction cracks in an Arctic tundra environment', *Arctic* **25**, 142–50

KUENEN, P. H. (1953), 'Significant features of graded bedding', *Bull. Am. Ass. Petrol. Geol.* **37**, 1044–66

LACHENBRUCH, A. H. (1962), 'Mechanics of thermal contraction cracks and ice-wedge polygons in permafrost', *Geol. Soc. Am. Spec. Pap.* **70**, 69 pp.

LEFFINGWELL, E. K. (1915), 'Ground ice wedges', *J. Geol.* **23**, 635–54
 (1919), 'The Canning River region, northern Alaska', *U.S. geol. Surv. Prof. Pap.* **109**, 1–251

LUNDQVIST, J. (1962), 'Patterned ground and related frost phenomena in Sweden', *Sveriges geol. Unders.* **583**C, 101 pp.
 (1969), 'Earth and ice mounds: a terminological discussion' in PÉWÉ, T. L., *op. cit.* (1969), 203–15

McARTHUR, D. S. and ONESTI, L. J. (1970), 'Contorted structures in Pleistocene sediments near Lansing, Michigan', *Geogr. Annlr* **52**A, 186–93

MACKAY, J. R. (1962), 'Pingos of the Pleistocene Mackenzie River delta', *Geogr. Bull.* **18**, 21–63
 (1966), 'Segregated epigenetic ice and slumps in permafrost: Mackenzie delta area, N.W.T.', *Geogr. Bull.* **8**, 59–80
 (1968), 'Discussion of the theory of pingo formation by water expulsion in a region affected by subsidence', *J. Glaciol.* **7**, 346–51
 (1971), 'The origin of massive icy beds in permafrost, western Arctic coast, Canada', *Can. J. Earth Sci.* **8**, 397–422
 (1972), 'The world of underground ice', *Ann. Ass. Am. Geogr.* **62**, 1–22
 (1973), 'The growth of pingos, western Arctic coast, Canada', *Can. J. Earth Sci.* **10**, 979–1004

MELNIKOV, P. I. (1966), quoted in J. DEMEK (1968), 'Cryoplanation terraces in Yakutia', *Biul. Peryglac.* **17**, 91–116

MIDDENDORFF, A. T. VON (1867), see letter by A. CAILLEUX, *J. Glaciol.* **10** (1971), 411

MISENER, A. D. (1955), 'Heat flow and depth of permafrost at Resolute, Cornwallis Island, N.W.T., Canada', *Trans. Am. geophys. Un.* **36**, 1055–60

MITCHELL, G. F. (1971), 'Fossil pingos in the south of Ireland', *Nature, Lond.* **230**, 43–4

MORGAN, A. V. (1971), 'Polygonal patterned ground of Late Weichselian age in the area north and west of Wolverhampton, England', *Geogr. Annlr* **53**A, 146–56

MULLENDERS, W. and GULLENTOPS, F. (1969), 'The age of the pingos of Belgium' in PÉWÉ, T. L., *op. cit.* (1969), 321–35

MÜLLER, F. (1959), 'Beobachtungen über Pingos. Detailuntersuchungen in Ostgrönland und in der Kanadischen Arktis', *Meddr Grönland* **153** (3), 127 pp.

(1962), 'Analysis of some stratigraphic observations and radiocarbon dates from two pingos in the Mackenzie Delta area, N.W.T.', *Arctic* **15**, 278–88

MULLER, S. W. (1947), *Permafrost or permanently frozen ground and related engineering problems* (Ann Arbor), 231 pp.

PALMER, A. C. (1967), 'Ice lensing, thermal diffusion and water migration in freezing soil', *J. Glaciol.* **6**, 681–94

PENNER, E. (1960), 'The importance of freezing rates in frost action in soils', *Proc. Am. Soc. Test. Mater.* **60**, 1151–65

PÉWÉ, T. L. (1959), 'Sand-wedge polygons (tesselations) in the McMurdo Sound region, Antarctica', *Am. J. Sci.* **257**, 542–52

(1962), 'Ice wedges in permafrost, lower Yukon river area, near Galena, Alaska', *Biul. Peryglac.* **11**, 65–76

(1966), 'Paleoclimatic significance of fossil ice wedges', *Biul. Peryglac.* **15**, 65–73

(1969) (ed.), *The periglacial environment* (Montreal)

PIHLAINEN, J. A. (1962), 'An approximation of probable permafrost occurrence', *Arctic* **15**, 151–4

PISSART, A. (1958), 'Les depressions fermées de la région parisienne: le problème de leur origine', *Revue Geómorph. dyn.* **9**, 73–83

(1963), 'Les traces de "pingos" du Pays de Galles (Grande-Bretagne) et du Plateau des Hautes Fagnes (Belgique)', *Z. Geomorph.* NF **7**, 147–65

(1967), 'Les pingos de l'île Prince-Patrick', *Geogr. Bull.* **9**, 189–217

(1968), 'Pingos, Pleistocene' in *The encyclopaedia of geomorphology* (ed. R. W. FAIRBRIDGE), 847–8

POPOV, A. I. (1956), 'Le thermokarst', *Biul. Peryglac.* **4**, 319–30

(1969), 'Underground ice in the Quaternary deposits of the Yana–Indigirka lowland as a genetic and stratigraphic indicator' in PÉWÉ, T. L., *op. cit.* (1969), 55–64

PORSILD, A. E. (1938), 'Earth mounds in unclaciated Arctic northwestern America', *Geogrl Rev.* **28**, 46–58

PRICE, L. W. (1972), 'Up-heaved blocks: a curious feature of instability in the tundra', *Icefield Ranges Res. Proj. Scient. Results* (ed. V. C. BUSHNELL, and R. H. RAGLE, *Am. Geogr. Soc.& Arct. Inst. N. Am.*) **3**, 177–81

RAPP, A. and ANNERSTEN, L. (1969), 'Permafrost and tundra polygons in north Sweden' in *The periglacial environment* (ed. PÉWÉ, T. L.), Montreal, 65–91

RAUP, H. M. (1965), 'The structure and development of turf hummocks in the Mesters Vig district, north-east Greenland', *Meddr Grønland* **166** (3), 112 pp.

RAY, L. L. (1951), 'Permafrost', *Arctic* **4**, 196–203

REFFAY, A. (1964), 'Quelques observations sur les buttes gazonnées des pâturages de la région de Saint-Claude (Jura)', *Revue Géogr. alp.* **52**, 315–23

RITCHIE, A. M. (1953), 'The erosional origin of the Mima Mounds of south-west Washington', *J. Geol.* **61**, 41–50

SCHAFER, J. P. (1949), 'Some periglacial features in central Montana', *J. Geol.* **57**, 154–74

SEPPÄLÄ, M. (1972a), 'The term "palsa"', *Z. Geomorph.* **16**, 463

(1972b), 'Pingo-like remnants in the Peltojärvi area of Finnish Lapland', *Geogr. Annlr* **54**A, 38–45

SHARP, R. P. (1942a), 'Periglacial involutions in north-eastern Illinois', *J. Geol.* **50**, 113–33

(1942b), 'Soil structures in the St. Elias Range, Yukon Territory', *J. Geomorph.* **5**, 274–301

(1942c), 'Ground-ice mounds in tundra', *Geogrl Rev.* **32**, 417–23

SHEARER, J. M. *et al.* (1971), 'Submarine pingos in the Beaufort Sea', *Science* **174**, 816–18

SHUMSKIY, P. A. (1959), quoted by J. R. MACKAY (1966)

SIGAFOOS, R. S. and HOPKINS, D. M. (1951), 'Frost-heaved tussocks in Massachusetts', *Am. J. Sci.* **252**, 55–9

STEARNS, S. R. (1966), 'Permafrost (perennially frozen ground)' *U.S. Army Cold Reg. Res. Engng Lab. Tech. Rep.*, **1**(A2), 77 pp.

SPARKS, B. W., WILLIAMS, R. B. G. and BELL, F. G. (1972), 'Presumed ground-ice depressions in East Anglia', *Proc. R. Soc.* **327**A, 329–43

SVENSSON, H. (1969), 'A type of circular lake in northernmost Norway', *Geogr. Annlr* **51**A, 1–12

(1970), 'Termokarst', *Svensk geogr. Årsbok* **46**, 114–26

SYNGE, F. M. (1956), 'The glaciation of north-east Scotland', *Scott. geogr. Mag.* **72**, 129–43

TABER, S. (1929), 'Frost heaving', *J. Geol.* **37**, 428–61

(1930), 'The mechanics of frost heaving', *J. Geol.* **38**, 303–17

(1943), 'Perennially frozen ground in Alaska: its origin and history', *Bull. geol. Soc. Am.* **54**, 1433–548

(1952), 'Geology, soil mechanics and botany', *Science* **115**, 713–14

TROLL, C. (1944), 'Strukturböden, Solifluktion, und Frostklimate der Erde', *Geol. Rdsch.* **34**, 545–694 (English translation, *Snow Ice Permafrost Res. Establ.*)

WALLACE, R. E. (1948), 'Cave-in lakes in the Nabesna, Chisana, and Tanana river valleys, eastern Alaska', *J. Geol.* **56**, 171–81

WARD, W. H. (1948), *J. Glaciol.* **1**, 146

WASHBURN, A. L. (1956), 'Classification of patterned ground and review of suggested origins', *Bull. geol. Soc. Am.* **67**, 823–66

(1969), 'Weathering, frost action and patterned ground in the Mesters Vig district, north-east Greenland', *Meddr Grønland* **176**(4), 303 pp.

(1973), *Periglacial processes and environments*

WASHBURN, A. L., SMITH, D. D. and GODDARD, R. H. (1963), 'Frost cracking in a middle-latitude climate', *Biul. Peryglac.* **12**, 175–89

WATSON, E. (1965), 'Periglacial structures in the Aberystwyth region of central Wales', *Proc. Geol. Ass.* **76**, 443–62

    (1972), 'Pingos of Cardiganshire and the latest ice limit', *Nature, Lond.* **236**, 343–4

WERENSKJOLD, W. (1953), 'The extent of frozen ground under the sea bottom and glacier beds', *J. Glaciol.* **2**, 197–200

WHITE, S. E., CLARK, G. M. and RAPP, A. (1969), 'Palsa localities in Padjelanta National Park, Swedish Lappland', *Geogr. Annlr* **51**A, 97–103

WIEGAND, G. (1965), 'Fossile Pingos in Mitteleuropa', *Würzb. geogr. Arb.* **16**, 1–152

WILLIAMS, R. B. G. (1969), 'Permafrost and temperature conditions in England during the last glacial period' in PÉWÉ, T. L., *op. cit.* (1969), 399–410

WOLFE, P. E. (1953), 'Periglacial frost-thaw basins in New Jersey', *J. Geol.* **61**, 133–41

YEFIMOV, A. I. and DUKHIN, I. YE. (1968), 'Some permafrost thicknesses in the Arctic', *Polar Rec.* **14**, 68

Proc., Permafrost Int. Conf., Indiana (Lafayette, 1963). *Nat. Res. Council Publ.* **1287**, 563 pp.

3

Patterned ground

When one steps upon the tundra almost anywhere in the (Lena) delta, it appears divided into countless irregular polygons of differing size, whose edges are higher than the middle. Between the edges of two such polygons, a small furrow is found, which is used as a pathway by lemming. (A. VON BUNGE, 1884)

One of the most conspicuous and characteristic features of periglacial regions is patterned ground, first described over a century ago. It occurs almost entirely within the cold-climate zones, although some minor forms of patterned ground are exceptional in this respect. The features that show marked sorting are certainly restricted to a climate where temperatures frequently fall below freezing-point, and both frost and the formation of ground ice are fundamental factors in the origin of patterned ground. A wide variety of forms comes under the heading of patterned ground; it is necessary, therefore, to classify the forms before their characteristics can be adequately described and the processes of formation are discussed.

Patterned ground occurs in both an active state and a fossil state. In the latter case, it provides a useful indication of past climate. Fossil patterned ground has been recognized ever more widely as aerial photographs have become available for increasingly large areas, and such photographs are often the best means whereby it can be located. Patterned ground has been recognized on aerial photographs of parts of the south and Midlands of England (F. W. Shotton, 1960). It is revealed through minor changes in the soil often not apparent from ground level, which affect the density of crops or type of vegetation.

The two most important aspects of the patterns are their shape and the degree of sorting that has developed. Classifications must, therefore, recognize these two elements, and they are the basis of a classification suggested by A. L. Washburn (1956). This purely descriptive classification has advantages in the present state of our knowledge concerning the origin of many of the features. Washburn bases the primary division of his classification on the shape of the pattern. A subdivision of each shape is then based on the character of the sorting, so that the complete classification is as follows:

Circles — sorted and unsorted
Nets   — sorted and unsorted
Polygons— sorted and unsorted
Steps   — sorted and unsorted
Stripes — sorted and unsorted

The features may grade into one another. Elongated steps or circles may be called garlands, where they are outlined by larger stones, and these merge into stripes as they become increasingly elongated. The circles form where the operating processes can work in isolation from one centre. But where two or more centres can react with one another, the net or polygonal form will occur. The irregular mesh pattern of the net is intermediate between the circles and polygons, which have angular and often regular boundaries. Some of the unsorted forms are made conspicuous on the surface by changes in vegetation or elevation. Earth hummocks, described in Chapter 2, are a common form of unsorted nets. At times, these features are in the form of bare patches surrounded by vegetation.

The gradient of the surface is the essential factor that determines whether a circle, net or polygon will form as opposed to steps and stripes. The circles, nets and polygons are restricted to flat or nearly flat surfaces. The steps and particularly the stripes occur on ground with a considerable degree of slope, varying between 6° and about 15–30°. Slopes of 2–6° have forms that are intermediate between the truly circular or polygonal forms and the true stripes. Slopes steeper than 30° are usually too steep to develop patterns.

More recently, Washburn (1970) has suggested a genetic approach to the classification of patterned ground. This is based on the following premises:

1 patterned ground is polygenetic
2 similar forms may result from different processes
3 the same processes may produce different forms
4 there are more processes than those currently recognized
5 the terminology should be simple.

Table 3.1 only gives the initial form, but this can be modified over time. For example, cracking produces a non-sorted (N) form, but this may later develop into a sorted (S) form. For some features (see Table 3.1) cracking is essential, and this produces polygonal forms, although the polygons can be deformed to become nets. The classification is based on both shape and process. Thermal contraction, which is an important process, can be caused by salt, seasonal frost or permafrost. The process of cracking is not essential for the development of sorted types.

1 Description of patterns

Each of the main types of pattern will be briefly described and then some of the more important forms will be examined in greater detail in an attempt to arrive at conclusions concerning their method of formation.

Table 3.1 Genetic and morphological classification of patterns (Washburn, 1970)

Pattern		Desiccation	Dilation	Salt	Seasonal frost	Permafrost	Frost action along joints	Primary frost sorting	Mass displacement	Differential frost heaving	Salt heaving	Differential thawing and eluviation	Differential mass wasting	Rillwork
		Cracking essential						Cracking not essential						
Circles	N								*	*	*			
	S						*	*	*	*	*	*?		
Polygons	N	*	*	*	*	*	*	*?	*?					
	S	*	*	*	*	*?	*	*?	*?	*?	*?			
Nets	N		*		*	*?		*?	*	*				
	S	*	*		*	*?		*	*	*	*	*		
Steps	N							*	*?	*?			*?	
	S							*	*	*	*	*?	*	
Stripes	N	*?	*?		*?	*?	*?		*	*	*		*?	*?
	S	*	*		*?	*?	*	*?	*	*	*	*?	*	*

S=sorted, N=non-sorted

(*a*) *Circles* can occur singly or in groups. In the sorted type there is normally a border of coarser stones around an area of finer material (Plate IX). Circles of this type are well developed in Spitsbergen where they attain dimensions of 0·8–3 m in diameter. The stones in the border appear to get larger as the size and thickness of the border increase. There are also good examples in west Baffin Island, particularly where limestone outcrops and forms a litter of small, flat stones, or on raised beach material. These features form in many polar and high mountain areas, and are not restricted to areas of permafrost. They occur in Iceland, for instance, in areas where there is no permafrost. Unsorted circles can also occur even in environments where no frost occurs, for example in parts of Australia.

(*b*) *Nets* occur where the pattern is less circular but not yet polygonal. Earth hummocks of unsorted material come in this category. Again, permafrost is not necessary. Nets occur in parts of Scandinavia, Siberia, Iceland and Canada, and are most common in sub-arctic and Alpine areas. The features are usually fairly small, those in Iceland being 1–2 m in diameter.

(c) *Polygons* are the best known type of patterned ground and their description and classification has given rise to a wide variety of terms. Polygons can form either in permafrost areas or in areas having seasonal frosts. Indeed, some types of polygon can form in warm semi-arid environments, where they are the result of desiccation. If sorting is present in this instance, it may be the result of stones being washed into the cracks by seasonal floods.

There are types of polygon that only form where there is permafrost. These forms are particularly valuable for palaeoclimatic studies. The type of polygon requiring permafrost must be differentiated from the much smaller features that can be seen at higher levels on the hills of Britain, for example, on the top of Helvellyn. These small polygons have been shown to be growing actively under present climatic conditions in Britain (S. E. Hollingworth, 1934). In the Tinto Hills in Scotland (R. Miller *et al.*, 1954), they have been destroyed purposely, yet have reformed in a few years.

The truly arctic features, which only form where permafrost is present, have been called tundra polygons or ice-wedge polygons. The former term has been preferred by many as it does not imply a specific formative mechanism. However, the association of the large tundra polygons with ice wedges is now very clearly established, and their method of formation is fairly well agreed, as outlined in the previous chapter.

T. L. Péwé (1963) has discussed the climatic conditions necessary to generate the ice wedges that form the polygonal pattern. He divides the ice wedges into three categories and the same classification can be applied to the polygonal forms that are the surface manifestation of the ice wedges. Active ice wedges are associated with the zone of continuous permafrost where temperatures fall below $-6°C$ and sometimes to $-12°C$. The mean annual number of degree-days of freezing ($°C$) ranges from 2800–4500. In the areas where inactive ice wedges occur, the temperature conditions are less severe, with mean annual values ranging from $-2°C$ to about $-6°C$ to $-8°C$, and 1700–4000 degree-days of freezing. The zone where these conditions occur lies to the south of the active ice-wedge zone. Permafrost is discontinuous, while frost cracking rarely occurs and ice is not added to the wedges, although the climate is cold enough to preserve them. Péwé's third category comprises fossil ice wedges which occur where the ice has melted and been replaced by sediment. Those found in central Britain imply a fall of temperature of 16–18°C below the present.

The true ice-wedge type of tundra polygon is considerably larger than the small type which develops by frost sorting of material. The tundra type may be over 10 m across, and those in the Avon valley in England (Shotton, 1960) range from a minimum of 4·5–6 m to a maximum of 60 m. The fossil ice-wedge polygons on the North Yorkshire

Plate VIII (*above opposite*)
Sorted polygons, 2–5 m in diameter, Devon Island, Canada. (C.E.)

Plate IX (*below opposite*)
Stone circles, Foley Island, Baffin Island. (C.A.M.K.)

Moors near Scarborough are between 15 m and 30–40 m across. The smaller forms on the other hand are often not larger than about 3 m and those described by Hollingworth (1934) in the Lake district are only about 0·6 m across. Size provides a fairly reliable method of distinguishing the two types of very different polygons. The smaller ones are also nearly always of the sorted variety. The larger ones, on the other hand, are basically unsorted, although in the fossil type the pattern is made by the sedimentary fill of the fossil ice wedges. The polygonal patterns are on the whole more common than the circles or nets, as where conditions are suitable for the formation of one, they usually allow many to form and thus the polygonal pattern develops. The tundra polygons sometimes develop a tetragonal or rectangular pattern and sometimes the five- or six-sided polygonal form. The polygons are nearly always formed in sediments but on rare occasions polygonal patterns have arisen where intense frost activity, working along polygonal jointing in bedrock, has given rise to these patterns in bedrock. Examples of frost cracks in the Palaeozoic limestone on Foley Island in Foxe Basin (Baffin Island) illustrate this point. The limestone in this area shows extreme frost shattering.

J. V. Drew and J. C. F. Tedrow (1962) have classified arctic soils and related the soils to patterned ground features. In Lithosols and Arctic Brown soils, both of which are well drained, the sorted type of polygon occurs. Non-sorted tundra or ice-wedge polygons occur in Arctic Brown and impeded drainage soils, which include Upland Tundra, Meadow Tundra and Bog soils. Tundra polygons are divided into types A to F, in order of increasing relief. Type F forms mounds or ridges, type E have fairly high raised rims, type D have lower raised rims, type C have slightly raised rims, type B have troughs only, and type A have very small troughs only. In Arctic Brown soils only types A and B occur. In Upland Tundra soils, types A, B and F occur and in Meadow Tundra soils, types B, C, D, E, and F occur, while in Bog Tundra soils, types C and F occur. The widths of the ice wedges increase in types D to F but are more or less constant in types A to C, although in these types the depth of the ice wedges increase from A to C. Thus the greatest relief of the ridges is associated with the largest ice wedges, which occur in the soils in which drainage is most impeded.

H. Svensson (1969) describes polygons on the distal side of a delta spread in north Norway now at 72–3 m above sea level. The polygons are outlined by shallow furrows; in summer some had rough-walled open fissures about 5 mm wide in the centres of the furrows. They could be desiccation features resulting from drying out of the peat in the summer and thus secondary phenomena not associated with ice-wedge formation. They do not penetrate into the mineral soil below. In late winter a greater area showed sharp, smooth-walled fissures, which also occurred in water-filled furrows and did not extend beyond them. The winter fissures were clearly not caused by desiccation and the mineral soil was seen to contain ice veins. The area is one of marginal permafrost, and in summer no frozen ground was found to a depth of 1·5 m. The polygons are not, then, formed by recent permafrost, but the fissuring is nevertheless a recent frost process. They may be termed frost-crack polygons, rather than ice-wedge polygons, and the process forms both the furrows and the crack by repeated thermal contraction. Lateral displacement of the soil leads eventually to the formation of the furrow.

Their distribution is affected by exposure, and they occurred where the snow cover was minimal.

A. Pissart (1968) has studied currently active polygons of the frost-fissure and ice-wedge type in Prince Patrick Island in the Canadian Arctic. The polygons are formed by ice, mineral soil, or sand and ice wedges, according to the soil type and summer humidity. The ice-wedge polygons occur on low slopes where moisture is abundant. The sand- and ice-wedge types may measure 5–30 m across and are often rectangular, occurring on broad crests in sandy material. The purely mineral wedges are found in very dry areas of sand soil.

Active patterned ground on the Avalon Peninsula of Newfoundland has been investigated by E. P. Henderson (1968). The features resemble those formed under permafrost conditions although there is no permafrost in the area at present. The patterned ground forms as a result of low summer temperatures due to fog, thin or absent snow and a strongly maritime climate. The patterns, which occur in hard till, include sorted polygons and circles as well as non-sorted circles and debris islands. Polygons are 1·8 to 3·6 m across and are transitional between large tundra forms and smaller, miniature ones. The climate provides abundant moisture and there is little reserve of heat in the soil after summer. The number of thaw-freeze cycles is greater than in a more continental climate, and freezing is the major driving force creating the patterns.

(*d*) *Steps* form where the slope is rather steeper. In the sorted type, the step is usually bordered by embankments of stones larger than the rest of the material. The steps sometimes form parallel with the slope contours, but at times they become drawn out into lobate form, elongated down-slope, forms which are referred to as stone garlands. Good examples occur on the slopes of the Glyders in north Wales. The term stone-banked terrace has also been applied to these features. They do not normally have the regularity characteristic of polygons although the sorted steps and garlands normally occur in groups. Tabular stones at the margins of the features are often arranged vertically in the lateral edges, but dip 60–70° upslope at the lobate front.

In the unsorted variety of this form, vegetational changes often show up the position of the features. Generally the riser of the step is well vegetated and the tread is bare. For this reason, the terms 'turf-banked terrace' or 'turf garlands' have been used to describe the features. In shape they resemble the sorted varieties, but they are often of smaller dimensions.

Neither of these forms need necessarily be associated with permafrost. Since their features are also strongly affected by mass movement, they will be considered in more detail in Chapter 4.

The somewhat irregular step and garland forms, and also polygons, merge into *stripes* as the slope increases. The sorted stripes are elongated down the slope and consist of alternating strips of coarser and finer stones. Again the dimensions vary very much; at one end of the scale is the small type that occurs in areas of seasonal frost activity. R. Miller *et al.* (1954) have described well developed small stripes in the Tinto Hills where the slope is about 20°. These stripes are forming actively under present climatic conditions. The coarse stripe is from 20–35 cm in width. Coarse material collects in a furrow while finer material forms a slight ridge. The depth of the furrow is about 7·5 cm. In better developed examples the coarse stripe may be 1·5 m wide and the finer two to four times the width. In places the features are quite extensive and have been traced for over 100 m down the slope under favourable conditions. The depth to which the stripes penetrate correlates on the whole with their size, the two increasing together. Because the stripes form on fairly steep slopes, they have less chance of survival when conditions suitable to their formation cease. They are not, therefore, often preserved in a fossil form.

In the unsorted type, vegetation again outlines the pattern and they have been termed 'vegetation stripes'. One report of these features from Arctic Canada describes vegetation in slight troughs 0·3–0·6 m wide and 3–4·5 m apart. But in other areas the widths of the vegetated and bare strips are about equal. This form is not necessarily confined to cold climates and has been reported from parts of Australia.

2 The processes forming patterned ground

In discussing the method of formation of patterned ground, the polygonal pattern will be considered in most detail. Circles, nets and polygons are formed by similar mechanisms and the same processes operate on steeper slopes to produce steps and stripes. Many different possible processes have been suggested. In order to assess their validity, detailed observations of the features and the changes occurring within them, as well as measurements of the processes and controlling factors, are necessary. Detailed studies of the polygonal features will be mentioned first, and then experiments, both in the field and laboratory, on the movements of material and the forces involved will be described. The matter is complicated by the fact that the features can form in a variety of environments by different processes and yet possess many of the same characteristics.

Definitions In discussing the formation of patterned ground the concept of the *active layer* is a useful one. As explained in the last chapter, this is the layer which melts during the summer season in an area where the ground is permanently frozen at depth. It is, therefore, in this layer that material can be moved and the patterned ground can form as the material is subject to alternating freezing and thawing. The term *frost-susceptible* is applied to material of particular grain size composition (see p. 34 and Fig. 2.5) that is subject to frost-heaving. A *non-frost-susceptible* soil is one which consists of coarser sediments with less than 1 per cent fines, the soil not being subject to frost-heaving. The term *fines* is applied to particles of less than 0·074 mm diameter, or passing through sieve 200 ASA. This includes silt and clay particles. The coarser material consists of sand, gravel and stones.

2.1   *Regional studies and observations of polygons*

As already mentioned, the polygonal patterns include two very different types of feature. The ice-wedge polygons are not necessarily associated with sorting of material.

On the other hand, some polygons are essentially sorted features and the movement of the material that causes the sorting is an essential part of their formation. In dealing with the development of this type of feature, it is essential to study the pattern both on the surface and in depth. The study in depth is not easy in an area of permafrost and A. E. Corte (1963) considers that a bulldozer is an essential piece of equipment in the study of these features.

(a) *North-west Greenland*   Corte has described the sorted polygons and other forms of patterned ground in the Thule area of Greenland, where they are formed on outwash. The features vary from very small centres of fines, only a few centimetres across, to large features up to 3 m in diameter. Some of the centres are raised, others are depressed; some are isolated and others are closely grouped. The fine material in the centres of the polygons consists of fines and sand in places, but elsewhere it is composed of small pebbles. The sorted features do not occur in the same areas as the tundra polygons. The tundra polygons tend to disappear or thin to very narrow cracks as they enter an area of sorted polygonal features.

The area chosen for close study by Corte included both types of polygon and also the junction zone between them. It was 80 × 60 m in size and at the northern and southern margins contained tundra polygons. A depression of washed coarser material in the centre contained the sorted features. The ice wedges were apparent as troughs into which some of the large material had been washed from the immediate vicinity. Trenches were bulldozed across the whole area to reveal the structure of the active layer and to study the nature of the ice wedges. In the area of the ice wedges, the active layer showed no disturbance and the horizontal stratification of the original deposits was preserved. Beneath the active layer, the sand and gravel was frozen and solid ice was found only in the ice wedges. When the profile was traced into the sorted area, the stratification was disturbed and amorphous ice masses appeared within the material. The proportion of fines increased and the surface became sorted. Where the surface material was finer, a plug of fine material was found beneath the surface. Within the plug the elongated stones had their long axes vertical, while in the permafrost beneath there were ice masses containing layers of fine particles and odd stones. The ice masses were covered by layers of fines. It appears, therefore, that there is a difference in the character of material in the undisturbed active layer in the area of the ice wedges and in the contorted active layer in the area of sorted polygons.

In order to establish the reason for this difference, soil samples were taken from the disturbed, slightly disturbed and undisturbed parts of the active layer. The samples were taken from the top, middle and bottom of the active layer, while those from the disturbed section came from the plug of finer material. The results of the analysis showed that the disturbed layer consisted of considerably finer material than that in the undisturbed layer. There was only 2 per cent fines in the undisturbed layer and little difference between the top, middle and bottom samples. In the disturbed layer, there was 5 per cent fines at the top, 11 per cent fines in the middle and 14 per cent fines at the bottom. This distribution shows the effect of vertical sorting. An evaporite deposit was found on the lower side of the stones in the undisturbed active layer, but this was missing from the large stones in the disturbed layer.

Corte also studied the character of the ice associated with the polygonal patterns. In the ice-wedge or tundra polygons there were ice wedges below the troughs on the surface. The fabric of the ice revealed thermal contraction cracks along the axial plane of the wedge (Chapter 2). In these cracks, small crystals of ice, 1–3 mm in size, were orientated horizontally. On either side of the more recent crystals were older ones that were orientated vertically and were 2–3 mm long, dipping steeply towards the axial plane of the wedge. The crystals must grow and become reorientated, therefore, as the contraction stresses occur.

The ground-ice conditions under the sorted polygons were different. There were amorphous masses of ice, containing some contorted layers of fine material beneath the centres of fines. Ice wedges were much smaller, if they were present at all. The amorphous ice masses consisted of large-grained transparent ice, with crystals up to 10 cm² in size, where there were no silt layers. But where there were dirt or silt layers, the crystals were much smaller. The C-axes of crystals tended to lie at 25–45° to the vertical. The contact of the massive ice and the wedge ice showed that the wedge ice was the younger of the two types.

Corte describes another type of feature consisting, on the surface, of mounds and depressions of low relief, formed in fine grained outwash. This pattern does not show surface sorting, but sorting was well developed vertically. Plugs of finer material occurred, containing at times as much as 28 per cent fines at the bottom and 5 per cent at the top. The active layer showed considerable disturbance and only small remnants of stratification survived between the plugs of fine material. The soil in these intervening bands only contained 2 per cent fines. The ice in these areas consisted almost entirely of ice masses, which were covered by layers of fine material. There appeared to be no systematic relation between the ice features of the active layer and the pattern on the surface, which showed no sorting. This lack of sorting may have been related to the absence of coarser material in this area, which contained no boulders or cobbles.

It appears that the nature of the material plays an important part in the character and development of sorting. Corte indicates three types of material, each of which produces a different pattern. The coarsest material, in which the percentage of fines is less than 2 per cent, produces ice-wedge polygons, and there is little disturbance of the bedding or surface sorting. The material of intermediate particle size produces uneven ground with sorted polygons and fine centres, in which ice masses occur below the surface and the bedding is lost. In the finest material, dense ice masses form and patterns of elevation and depression develop, with relief up to 30 cm. There is little sorting by size. In the last two conditions, ice wedges do not form conspicuously. The sorting of the surface layer takes place most readily when the percentage of fines in the plugs is between 3 and 8 (Fig. 3.1).

Cyclic changes in patterned ground have been examined by J. W. Marr (1969) in the Thule district of Greenland. The fine particles in bouldery matrices form islands that are the tops of plugs of fines rising through the coarse active layer above the permafrost table. The 'islands' vary in area from 0·1–80 m², and in colour from light sand to almost black, while in profile, some are concave and some convex. The islands can be arranged in a dynamic sequence according to their vegetation. The succession starts with mosses and lichens, passing through luzula, willow and heather, to the mature phase of brownish heather over a convex substratum of lower plants. As the sequence disintegrates, cracks develop in the ground cover and the substratum becomes concave; the plants are eventually blown away, leaving lichen-poor gravel and cobbles in a matrix of lichen-rich boulders. The cyclic processes continue in a generally stable environment and are the product of internal processes in the debris island ecosystem. The building process is the emergence of fines, possibly owing to local ice formation in the active layer; the melting of this ice plug sets up the disintegration part of the cycle, the melting resulting from heat absorbed by the dark plants.

(*b*) *North-east Greenland* A useful summary of recent work on patterned ground in the Mesters Vig area is provided by Washburn (1969). The mean annual temperature is $-9 \cdot 7°C$ (mean of coldest month $-24 \cdot 3°C$, warmest month $5 \cdot 9°C$). The mean annual precipitation is $372 \cdot 5$ mm, most of which falls as snow. *Non-sorted circles* occur widely, and have bare centres surrounded by tundra plants. On a 1° slope they are elongated

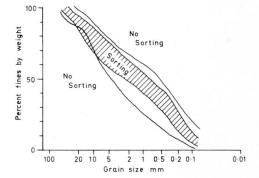

Fig. 3.1
Relationship between size of material and degree of sorting in periglacial deposits (A. E. Corte, *Biul. Peryglac.*, 1963)

down-slope. Their diameters are 1–1·8 m, the central area being up-domed by 11–12 cm. Sub-surface observations show contorted material in the moister borders. Mass displacement of the underlying fine material into the overlying coarser debris probably takes place. Downward movement takes place in the borders while upward movement affects the centres which are underlain by finer material. These movements are probably caused by moisture-controlled changes in intergranular pressure. *Small non-sorted polygons* are less than 1 m across and can occur in the centres of the sorted nets as cracks form, giving patterns of 15 × 20 cm to 20 × 45 cm. Some of these on delta deposits are undoubtedly desiccation cracks up to 15 cm deep and spaced 20–50 cm apart. They are formed in sand and the depth of the crack is related to the sand thickness. Dilation during freezing is a possible origin, but contraction is more likely for the polygonal forms. Desiccation could be caused by withdrawal of water to form ice, or by drainage and evaporation. These small polygons are confined to the active layer. *Large non-sorted polygons* on arkosic bedrock are outlined by polygonal systems of vegetated furrows, with a circular, rather bare centre. They attain 5·5–10·5 m in diameter, with either flat or domed centres. The furrows are up to 0·5 m deep. Bulldozed sections revealed unconsolidated material in the form of disintegrated bedrock *in situ*, but somewhat disturbed by frost. Dips increase towards the furrows, while curved fault lines indicate hinged zones of collapse. These structures clearly result from frost cracking, heaving and thawing. The furrows suggest melting of underlying ice masses. Similar features also formed in gravelly sand and till. *Small non-sorted stripes* less than 1 m across are probably caused by rill action; frost action may also contribute by causing micro-heaving of nubbin-like areas to concentrate the rills. In some areas, however, they are too regular in spacing to be related to rills, and may be caused by desiccation cracking. *Large non-sorted stripes* occurring in till and delineated by vegetation furrows probably originate by frost cracking and gelifluction.

Small sorted nets and polygons differ in shape, and since polygons and circles may form by different processes, the distinction is important. They are probably of comparable origin with the non-sorted forms, being basically desiccation features. The *large sorted nets and polygons* are the most common form and occur wherever there is a high moisture content throughout the season. Their borders are irregular and stony, the mesh elongated down-slope. Diameters range from 0·1–8 m, and the borders are 15–100 cm wide. Some centres are dome-shaped, while the borders are wedge-shaped with slabby stones standing on edge. Humic material occurred 15 cm below the surface adjacent to one central area, and moss was buried under a stone. The structure is consistent with the upward movement of silty clay and lateral surface spreading. Even on slopes of 1–2°, the nets are elongated down-slope. Frost cracking is responsible for the markedly polygonal forms, with mass displacement by frost action being manifest near the surface. Upward movement from depth is suggested by the dome shape, with fine material moving up into a rubbly cover. Changes in intergranular pressure consequent on freezing provide the most likely mechanism.

Small sorted stripes less than 1 m in width are uncommon. They are probably due to rill action in places but not everywhere. *Large sorted stripes* are much more frequent. Some occur on slopes as steep as 22°. Sandstone stripes alternated with dark trap rock debris in one area, and sorting was by size and lithology. Widths vary, being 0·3–1·0 m in stony areas and 1–3 m in till-like debris. The frost table is found to lie at a higher level in the stony stripe than the finer one. Frost wedging and heaving along nearly vertical bedrock joints probably explain the large sorted pattern in the Nyhavn hills. The lithological sorting may be caused by the faster down-slope movement of the coarse material, as in the lobate forms; frost action would assist this differential movement. These stripes are a mass wasting feature, with faster movement by creep or gelifluction.

In conclusion, Washburn suggests that patterned ground is polygenetic in origin, with the following processes being important:

1 desiccation cracking for small non-sorted polygons
2 frost cracking for large ones
3 dilation cracking for some small non-sorted polygons
4 sedimentation for some non-sorted circles
5 mass displacement for most non-sorted circles and large sorted nets, turf hummocks and large sorted stripes
6 rill action for some non-sorted stripes and small sorted stripes
7 differential mass wasting for large sorted stripes
8 frost wedging and heaving for other large sorted stripes.

(c) *Iceland*   Some of the periglacial patterns in Iceland have been described by J. D. Friedman *et al.* (1971). The major type they consider is the large-scale polygon of frost-crack type; these are forming currently without permafrost. Frost-cracking appears to have increased in the last 10 years. Tephrochronology has also enabled dates of more intense activity in the 1600s and 1700s to be established. The large-scale polygons are mainly concentrated in the centre of the country, more than 650 m above sea level. Where the subsoil is loessial, they occur down to 300 m. The tephra layers often bend down towards the polygonal furrows at depths up to 1 m, but where they are parallel to the surface the cracks must have formed before the tephra was deposited. Infrared aerial photography clearly reveals the patterns, which are tetragonal in eastern Doma-dalshals. These features are not genuine ice-wedge polygons and are not related to permafrost. The wedges are narrow and have only been formed fairly recently. Open cracks frequently occur suggesting active formation. The mean annual temperature above 600 m is probably less than $-1°$C, but not as low as the temperatures of $-6°$ to $-8°$C which are usually considered necessary for ice-wedge formation. The permeability of the inland desert areas of Iceland may help the cracking process.

(d) *Fennoscandia*   Ice-wedge polygons in Sweden have been traced on air photographs (H. Källender, 1967; A. Maack, 1967; S. Ohrngren, 1967). In the Padjelanta area where the polygons occur the mean annual temperature is about $-3°$ to $-4°$C. The area was deglaciated about 7000–8000 BP. The rather irregular polygons are usually about 10–40 m across and show up because of the denser vegetation in the sheltered troughs. The trenches are 0·2–2·8 m across and their depth is 0·05–0·4 m. In August the permafrost table lay at a depth of 0·5–1 m and no ice wedges or ice-wedge casts were revealed as wide as the surface trench, although thin cracks, 1 cm wide and about 70 cm long, were seen. Ground temperatures showed little annual variation below 2 m where many values were between $-1°$ and $-2°$C. Temperatures at the present are unlikely to be low enough in this area for contraction cracking to occur, and probably never have been low enough since deglaciation, so some other process must have been operative. Their spacing does not suggest an origin as desiccation cracks. Weak ice wedging and contraction cracking could have initiated the pattern in the past, but the increase in size of the troughs could be aided by running water, the wedging action of roots, wind erosion and a filling by lateral compression of once-wider frost cracks. Frost-heave in general could have produced the large-scale pattern. The pattern of large polygons is probably now mainly a fossil feature.

A. Rapp and G. M. Clark (1971) have described large unsorted polygons in Swedish Lapland at 67° 16′N, 16° 52′E. In these polygons which are formed on silty till, with crack fillings less than several millimetres thick, the sub-surface structures are not distinct, but it is thought that the polygons may be frost-crack contraction features. Wedge-shaped casts are much better developed on sandy outwash where they may be more than several centimetres thick: these are true tundra polygons originally developed on permafrost. Both types formed after the major deglaciation in 8000–9500 BP and are now largely fossil or inactive. Frost cracking occurred between 1963 and 1967, possibly in the hard winter of 1965–66 when the snow cover was thin. Measurements of temperature at depths down to 4 m showed perennial frost.

Polygonal features in Finnmark have been examined on air photographs and in the field by H. Svensson (1967). The area is above the forest line and the 0°C isotherm runs through it. Tundra polygons occur mainly on deltas, marine terraces and beaches, but also on Eocambrian bedrock and sediments. The pattern is orthogonal with sides 12 m long on Mount Bugtkjölen, where the pattern covers a large area of blockfield. The blocks, sandstone and conglomerate, range in size from boulders to coarse gravel. The pattern consists of furrows 15–25 cm deep in some of which stones stand on end. The lichen cover suggests stability at the present day. The bedrock is more than 1·25 m down. The most persistent pattern of fissures runs north 50° east to north 65° east, and is probably related to the bedrock joints.

(e) *British Isles*   Polygons and stripes in the Cairngorm Mountains of Scotland are restricted to heights above 900 m on the plateau. R. B. King (1971) distinguishes between coarse and fine types according to the ratio of boulders to sand. Two types of stripes occur: a coarse variety consisting of both bare and vegetated stripes, and those comprising lines of boulders in a sandy matrix. There is a significant difference in stone size between the two types. Some movement of the stones was recorded. During freeze-thaw, needle ice develops, elongated stones are upturned, and others less than a given size may be raised bodily. The fine polygons form by movement of finer material away from large stationary boulders whose mean size was 71·4 cm. Polygon size was found to be proportional to the spacing of large stationary boulders. The greater the amount of heaving, the larger the stationary boulders and the larger the size of the polygon. Coarse polygons are probably formed by selective weathering, as there is no evidence of movement. Sorting in this case may be related to a greater decrease in the size of small boulders since these are more effectively broken by frost shattering. The stripes probably formed by laminar downslope movement as many of the stones in the stripes have a low inclination. Gelifluction and frost creep are probably the main processes. The age of the coarse polygons and stripes was assessed by lichenometry, using a growth rate of 16 mm/century: they are probably of the order of 100–200 years old on this basis.

The variety of periglacial features found in northern England and described by L. Tufnell (1969) includes several examples of patterned ground. Periglacial action has taken place at various times, some features dating from the Weichselian, some from the Late-glacial, and some are currently active. The features that could be regarded as patterned ground phenomena include stone polygons varying from 1–15 cm in size, the larger ones being fossil; stone stripes, those currently forming being 7–10 cm wide, while again the larger ones are fossil and only occur above 600 m; erected stones caused by cryoturbation, and frost-patterned vegetated areas.

In north Wales, D. F. Ball and R. Goodier (1970) have shown that sorting has occurred and still occurs in Snowdonia where the climatic range is greatest. Small sorted polygons are found at 945 m near Foel Grath Carneddau. Table 3.2 shows a tentative scheme relating form to process. Large sorted stripes occur at heights over 450 m, with a width of 5–8 m and stones 60–150 cm long orientated down-slope. Small sorted stripes have repeat distances of 45–75 cm and occur on 25–30° slopes in the Rhinog Mountains and on Y Garn. Small sorted polygons have a size of 30–45 cm,

Table 3.2 Relationships between form and process for patterned ground in North Wales (Ball and Goodier, 1970)

	Dominant process		Time		
	Gelifluction	Cryoturbation	Late-glacial	Recent	Present
1 Stone-banked lobes	*	?	*	*	*
2 Turf-banked lobes	*		*	?	?
3 Terracettes	*		?	*	?
4 Large sorted stripes	*	*	*		
5 Small sorted stripes	*	*		*	*
6 Small unsorted stripes	*	*		*	?
7 Large sorted polygons		*	*		
8 Small sorted polygons		*		?	*
9 Earth hummocks		*		?	*

and observations in 1968–9 showed that a site at Foel Grach was frozen down to 20–30 cm from late November to March. Sorting and freezing were both limited to about 20 cm in depth. After the pattern was disturbed, it had started to reform after only one season.

(*f*) *Pyrenees* A detailed study has been made (K. Philberth, 1963) of the polygonal soils of part of the French Pyrenees. This is a middle latitude area without true permafrost, but one where the smaller, sorted features of the high mountain environment are well developed. Nearly all high mountains have examples of this type of polygonal pattern. The area studied includes mountains with summits over 3000 m, which is the altitude of the snow-line. Most of the polygons are small and show marked sorting, the edges of the polygons consisting of a mixture of material including many stones. In studying the distribution of the polygons it was found that they do not occur on the granite outcrops. Their absence could be accounted for by the lack of fine material of size less than 0·02 mm: water could, therefore, drain very readily from the soil. On the other hand, sorted polygons are well developed on schistose rocks, which weather to give a fine soil. The polygons develop best on well-weathered outcrops of schist in areas where snow-melt can add to the soil moisture. They only occur above 2700 m at elevations where small patches of soil can freeze from early autumn onwards.

Various factors influenced the size of the polygons, including the altitude, the depth of soil, the size of the stones and the stage of development. The size tends to diminish with increasing altitude and where the soil depth is small. Where soil thickness is not the limiting factor there is a tendency for large stones to form large polygons, but this is not always found. Until the polygons are fully formed and exert an influence on their neighbours their size is determined by the length of time they have been forming. At this stage they may be considered as embryonic forms.

Where the coarse border is formed of flat stones, these often stand on edge. The border of stones does not show any sign of soil on the surface, but by digging it was shown that the stones only extend 2–4 cm downwards. The lack of fine material resulted from the washing out of the coarse borders in this instance where the stones were small.

R. Miller *et al.* (1954) in their study of the polygons on the Tinto Hills in Scotland show that the sorting does not extend to a depth greater than 10–15 cm from the surface. Other observers have shown that the stone bands tend to narrow downward. T. T. Paterson (1940) showed that where the soil thickness over bedrock was 25 cm, the coarse stone borders extended throughout the soil thickness. H. W. Ahlmann, on the other hand, found polygons in which the stone borders widened downwards and extended to 60 cm which was the depth of the active layer in the area.

3 Field and laboratory experiments on polygons

Experiments on the movement of stones and the penetration of frost can give useful information with which the various theories of polygon formation may be tested. A. Pissart (1964) has carried out experiments on small polygons in the French Alps at Chambeyron. Two polygons, 80 cm in diameter, were destroyed in this area. In the first patch, small polygons, about 20 cm in size, reformed, and the outward movement of the stones was measurable. In the second patch, polygons did not reform. On another polygon, six pebbles were placed 15–25 cm from the centre of a polygon 160 cm wide. These six pebbles had moved 7, 7, 10, 15, 16 and 21 cm between 1947 and 1964. They were all found to be at the edge of small polygons about 10–20 cm across that lay within the larger one. The movement of the pebbles shows that small polygons can form under present-day conditions in this area at 2800 m. The small polygons are probably temporary and do not affect the movement of the pebbles in relation to the larger structure.

The formation of polygons such as these is closely related to the annual and diurnal temperature changes and the rate of frost penetration. In arctic regions, the annual freeze-thaw cycle is the most important (Chapter 1), but in high-altitude middle-latitude areas the daily freeze-thaw cycle is of greater significance. On clear days, the range at the surface in the latter environment may be 30–35°C, but it may be only a few degrees at a depth of 15 cm. In middle latitude areas, the depth of snow is more important than the mean winter temperature in determining the depth of frost penetration. Years of thin snow allow much deeper frost penetration than unusually cold years. P. J. Williams (1961) has suggested that the nearness of the mean annual temperature to 0°C, rather than the intensity of winter cold, determines the depth of frost penetration. The types of patterned ground that depend on deep frost penetration cannot form if the mean annual temperature is greater than 3°C. However, there are other forms of patterned ground that do not depend on this limit, such as stony earth circles.

Experiments have been carried out by Philberth (1964) to study the way in which frost penetrates and the hydrostatic pressure caused by freezing. For the experiments, he used a sample of soil that included both fines and stones, and that showed well developed natural polygons. The largest stones were 20 mm in size. Of the material below 2 mm in size, 60 per cent was below 0·2 mm and 30 per cent below 0·02 mm. Only those particles smaller than 2 mm were used for the experiments. The material was well mixed with 30 per cent water before the experiment started. A layer of the material was put in one vessel. In a second vessel, a thin layer of the material was covered with dry stones. Both vessels were subjected to freezing. The soil beneath the dry stones had frozen throughout, but in the first sample the frost had only penetrated 27 mm. At first, the soil froze in a relatively soft form, in which it could respond to external pressure because of its viscosity. However, as the temperature fell still further, the soil froze hard and solid.

In a second experiment with a lower rate of freezing, the frost penetrated only 6 mm into the first sample of uncovered soil, but the soil beneath the dry stones in the second sample had frozen to a depth of 56 mm. Thus moist, fine material freezes much less rapidly than coarse materials, and dry stony materials have a higher conductivity.

Experiments were carried out to measure the hydrostatic pressure of moist soil surrounded by a freezing border. Damp soil was placed in a cylindrical box open at the top, so that the soil could rise upward. An instrument to measure the hydrostatic pressure was placed in the centre of the box. It consisted of a compressible tube containing saline water, the level of the water indicating the pressure in the soil. The water in the tube started to rise after freezing had proceeded for 2–3 hours. During the following 9 hours, the water reached half its final height. Horizontal ice needles formed at the surface during the first few hours before the water started to rise in the tube; the rise took place after the surface appeared to have become impermeable. At first the surface was compressible, but later it became rigid. The maximum increase of water level in the tube indicated a change in volume of 2 cm³ in the box. The change in volume could have resulted from either pressure or expansion. Further experiments, using mercury in the tube, showed that at first the force was caused by pressure, but that in later stages, an increase in volume became important. This occurred when the surface ice layer became too strong and solid to give way to the pressure. The rise of the liquid was more rapid in the second phase.

Experiments were also carried out using lead boundaries between one large and two neighbouring small polygons. The experimental polygons were rectangular compartments filled with soil; the small polygons had a base of 70×70 mm and the larger 140×140 mm. The lead boundaries between the compartments were 2 mm thick, and were free to move except at their ends. The lead was cut into thin strips so that the movement at different depths could be studied. The damp soil filling the polygons was 70 mm thick, which gave a width to depth ratio of 1:1 in the small polygons and 2:1 in the large one, ratios suggested by field measurements. The rate of freezing was such that 80 mm of soil froze in 12 hours, the rate of frost penetration decreasing with time. As freezing took place, the lead bands between the polygons were pushed out towards the larger polygon from the smaller centres at a maximum rate at the bottom, the distance moved after one freezing period being 0·6 mm. At a depth of 60 mm, the movement was 0·2 mm. The values increased to 1·7 mm and 0·8 mm respectively and extended upwards to a depth of 50 mm (where the movement was 0·4 mm) after two freezing periods. After three freezing periods, the bending of the lead strips had penetrated to within 30 mm of the surface. The amplitude of movement was 2·3 mm at the base and 1·4 mm at 25 mm above the base. These values suggest that hydrostatic pressure plays an important part and that the pressure in the smaller freezing polygons is greater than that in the larger. This means that the smaller polygons will grow at the expense of the larger and this accounts for the uniformity in size that is found in any one small area, such as that studied by Philberth.

Corte (1962) has studied the differential movement of particles of different size in experimental conditions. He found that when freeze-thaw cycles operate in a horizontal plane, coarse particles will move up and fines will move down. Horizontal and vertical sorting will always occur together as a result of this differential movement. Vertical sorting is produced by freezing and thawing from the top, and horizontal sorting by freezing and thawing from the side. The degree of sorting is a function of the rate of freezing and gravitational forces. The finer fraction moves away from the cooling front, the fines moving downward when freezing takes place from the surface and laterally in a parabolic path when freezing is from the side. The finer particles migrate away from the coarser, and this causes the soil under the coarser ones to sink. The soil into which the finer particles move will heave. Maximum rates of freezing, measured in the field near Thule in Greenland, indicated a frost penetration of 12 mm/hour. The segregation of particles was greater with slow freezing taking place from the top downward.

Corte (1966) has recently discussed the results of detailed laboratory experiments on the effects of thawing and freezing. In summarizing his results, he states that the degree of sorting decreases as the amount of moisture decreases. The movement of particles depends on the amount of water between the ice-water interface and the particle, the rate of freezing, distribution of particles and the nature and position of the freeze-thaw plane. If a mixed soil is saturated and if the rate of freezing is slow, a large number of sizes can be sorted out. For example if a coarse heterogeneous soil is frozen it becomes better sorted if the rate of freezing is very slow, but if it is frozen very quickly, no sorting takes place. Fine particles can be sorted under a wide range of rates of freezing. The results of one experiment are shown in Fig. 3.2A. Some sorting occurred in the laboratory with a rapid freezing rate of 30–33 mm/hr, a rate which is higher than natural rates. However, lateral sorting under natural conditions should not be limited to a maximum rate of 12 mm/hr, at least for samples similar to those tested in the laboratory.

There are three mechanisms that allow sorting:

1 sorting by uplift or frost-heaving
2 sorting by migration in front of a freezing plane
3 mechanical sorting.

The first type occurs when freezing and thawing take place from the top; the second takes place when freezing and thawing proceed either from the top or sides. The third occurs when freezing is from the bottom and thawing from the top, and the process allows mounds to form. When the freezing and thawing occur from the top, the coarse particles move to the top and the fines move downward to the bottom of the freeze-thaw layer. This results in vertical sorting. When freezing and thawing occur from the sides, the particles move away from the cooling front and the coarsest remain on the cooling side. The result of this process is lateral sorting, Mechanical sorting occurs when mounds are formed and when the material freezes from the bottom upward. The presence of vertical sorting allows lateral sorting to take place if the direction of freezing changes. Pockets of fines can be a starting point in the vertical sorting process. Moreover, during vertical sorting, mounds can form and this assists the process of mechanical sorting as indicated in Fig. 3.2B. Both these effects are enhanced by lateral migration, which occurs if side freezing takes place. This can occur, for example, from the coarse stone borders that may penetrate downward to form a vertical plane and that possess a high conductivity.

The importance of the nature of the material in determining the ease with which

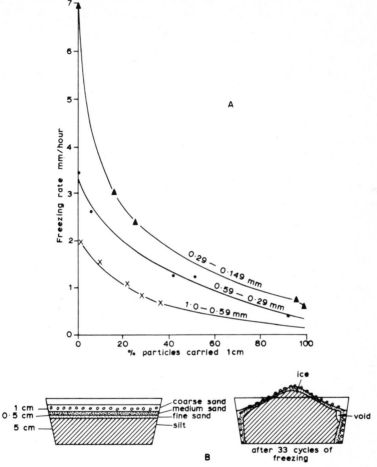

Fig. 3.2
Percentage by weight of quartz particles carried 1 cm in front of the freezing plane under different rates of freezing (A. E. Corte, *Research Report* 85, 1962, US Army C.R.R.E.L.)

polygons can form has been mentioned by S. Thorarinsson (1964) in describing polygons in Iceland. He has shown that polygons of the ice-wedge or tundra type are forming under present conditions in the fine loess soils in the interior of the country. In this type of soil, the polygons are about 20 m in diameter and occur down to levels of about 320 m in the north. Actively growing polygons in this type of soil probably only occur above 420–450 m above sea level (see p. 79). Those formed in coarser outwash soils only occur several hundred metres higher and are probably fossil. The polygons are about 15 m in diameter, with troughs 30–40 cm wide, which extend down as cracks to 80 cm. Sorted polygons of various dimensions also occur in Iceland.

One of the problems of studying patterned ground is the difficulty of obtaining undisturbed samples and sections in loose material. J. B. Benedict (1969) has described a technique for recording microfabrics by impregnating soil samples with plastic and measuring the grain orientation in thin section. Similarly, B. T. Bunting and R. H. Jackson (1970) have succeeded in taking samples of soil from sorted polygons on Devon Island, Arctic Canada, using a polyester resin for fixing. The physical properties of the material of a sorted polygon were examined using samples from different depths, both in the stone border and the clay plug. The permafrost table was 50 cm below the surface. The sorting of the clay plug was good, but the gravel margins showed no sorting. There were two contrasting types of sediment involved in the polygon, rather than the sorting of mixed material. There were calcareous efflorescences on the underside of larger flatter stones, indicating intense upward movement of carbonate-charged solution. The gravel of the border was coarse and angular at the surface and more rounded at depth; the gravel in the plug was generally coarse and fairly well rounded. The clay plug had a higher liquid limit and lower bulk density, and showed an accumulation of organic material in the centre. The dynamics of the system must relate to the interaction between the two different parts, the coarse and the fine. Some form of shear mixing or the setting-up of slip planes close to the steeply dipping junction may be involved. Convective mixing may also have a role to play.

4 Fossil Polygons

Although it is more common for fossil ice wedges and involutions to be preserved, fossilized patterned ground features have sometimes been recorded. Those found in parts of Britain, mainly eastern England, provide good examples. They have been studied both on aerial photographs and on the ground by R. B. G. Williams (1964) and by A. S. Watt *et al.* (1966). Their distribution is fairly widespread. Two major types of feature have been identified: first, the fossil ice-wedge polygons, and secondly the *trough* type, consisting of chains of polygons merging into stripes when the slope exceeds 2°. The trough type features make a large pattern with a mean polygon diameter of 9·4 m. They show up on the aerial photographs because the soil of the centres differs from that of the margins, leading to a difference of vegetation types. This is especially clear where the patterns are revealed by heather growing amongst grass. The pattern occurs where sandy and chalky patches alternate on Thetford Heath in East Anglia, where this type of trough polygon is well developed.

A trench on Thetford Heath revealed the subsurface pattern. Deep sand underlies the heather, filling ring-like or linear troughs, while the sand is very thin in the centres of polygons or stripes, which consist of very chalky till carrying a grass cover. The sand is probably Pleistocene and eolian in origin but the process by which it has reached the troughs is not clear. It is possible that this may have been by small-scale gelifluction or frost creep, moving a uniform sand cover into the troughs. The sand itself is not derived from the surrounding chalky material. The features show no evidence of sorting and can, therefore, be classified as unsorted polygons. They differ from the ice-wedge type in that the troughs are gently rounded and as wide as they are deep. The material beneath the troughs and centres, where it can be seen, is very contorted. The connection between the contortions and the troughs is not clear, however. The disturbed layer extends down to 1·5–1·8 m, indicating the thickness of the active layer during the formation of the features.

The trough polygons and stripes are restricted to the chalk outcrops of eastern England only, and do not occur on the more westerly chalk outcrops. The reason for this distribution is not clear, but it may be connected with the greater continentality of the climate in the east. It is likely that the features date from the last major glaciation, for they are restricted to areas outside its extent and they appear to have been little modified since.

In many other areas near the former limits of Pleistocene ice or in areas of severe cold in the Pleistocene, fossil patterned ground has been recorded. Most discoveries have come since the advent of aerial photography, for the surface patterns may be invisible or unnoticed in the low-angle ground view, especially in vegetated or cultivated areas. Fossil ice-wedge casts have been more frequently recorded than surface patterns since they can be very strikingly displayed in sections. Sometimes the discovery of ice-wedge casts will enable the associated polygonal surface markings to be identified. Fossil ice-wedge features are described and illustrated in Chapter 2. Some recent references to fossil-patterned ground include G. M. Clark (1968: Pennsylvania), J.-C. Dionne (1969: Quebec), A. Maack (1967: north Scandinavia), A. V. Morgan (1972: Ontario), S. Ohrngren (1967: north Norway), and H. Svensson (1967: south Sweden). Very well developed polygonal patterns have been located on air photographs in Provence east of the Rhône by A. Cailleux and C. Rousset (1968). They occur on gravels covered with limon which fills the cracks and was brought by the wind. The polygonal cracks are contraction features and imply a mean annual temperature of $-5°C$ or less. They are early Quaternary or early Riss in age. Their diameter varies from 10–30 m, the mode being 20–25 m. The associated fossil ice wedges are 2–3 m deep and 1·5 m wide at the top.

Late Weichselian polygonal patterns on till left by the Irish Sea ice occur northwest of Wolverhampton in central England (A. V. Morgan, 1971). The markings can be seen in vertical air photographs and ice wedges have been revealed in cuttings. 550 polygons were measured, varying from 50–150 cm. Some larger patterns up to 19·2 m were found with smaller ones inside them. The fill in the wedges consists of sand and some gravel, derived from the sediments overlying the till. Soviet workers have suggested that the larger polygons may form in milder conditions than the smaller ones, but this is unlikely to apply in Staffordshire, as the wedges do not penetrate below 2 m.

Black and T. E. Berg (1963) show how patterned ground can be used to record glacial fluctuations in Victoria Land, Antarctica. The patterned ground can be used to date ice advances, for the width of the ice wedges gives an indication of the time since they started to form. In this area, the wedges grow at a rate between 0·3 mm to more than 5 mm each year. All the patterned ground that occurs in this part of Antarctica could have developed during the last 10,000 years. In the older features, the raised rim is double, and this also gives an indication of age.

5 Processes forming polygons

Some of the processes causing polygons to form have already been mentioned in connection with polygon features. The main processes will now be discussed more systematically to provide a summary of present views on polygon formation. It is not possible to list all the hypotheses of polygon formation, for as J. K. Charlesworth (1957) has pointed out, there are almost as many views as there have been observers. All the different types of polygons mentioned cannot have been formed by the same processes. It is, therefore, necessary to divide the features into various categories. The two most important categories are the sorted polygons and the ice-wedge or tundra polygons.

5.1 The formation of sorted polygons

Philberth's views (1964) on the origin of sorted polygons have already been described. His theory is based on the effect of freezing pressure from the coarser border. This pressure causes the finer centre to bulge up, and as the pressure increases, the centre becomes smaller. The polygons, therefore, tend to become uniform in size, the smaller growing at the expense of the larger. The coarser border allows the cold to penetrate more readily and the border freezes before the fine plug, which is compressed by hydrostatic pressure. The stony borders are displaced towards the large polygons and the contact between two polygons causes them to become rectilinear. These suggestions are supported by the relationship between the size of the polygon and the depth of the active layer. The first stage in the process is the updoming of any fine centre that may exist in the soil. Permafrost is not required for the process to operate in mountain areas.

Washburn (1956) has assembled many of the earlier theories of polygon formation. He emphasizes the necessity of adopting a polygenetic origin to account for the wide variety of features and even for one type of polygon. Some of these theories will now be mentioned briefly.

Expansion caused by freezing was advocated as early as 1910 by B. Högbom. He suggested that the fines attract more water through capillarity. The fines expand more than the coarser parts, which are pushed out by the freezing centres of fines. Fine material is drawn back when the ground thaws and thus the material becomes segregated. Sorted polygons could form where the centres interact with each other.

A. Hamberg (1915) suggested a variation of this theory. He supposed that where stones overlie a fine layer, the fines would be heaved up where they approached the surface most closely. A vacuum would form beneath them into which more fines would be drawn until the fines reached the surface. The stones would then be moved off the upheaved centre to form a marginal circle or polygon.

S. Taber (1943) advocated differential heaving to account for polygons. He considered that differential heaving and gravity could cause sorting in mixed material. Local differential heaving is, however, common where there are no polygons and this process does not account for their even spacing. It has also been suggested that cryostatic pressure, caused by freezing downward towards the permafrost table, could result in heaving. Pockets of unfrozen material would remain and these might be pushed up through the already frozen crust or a coarser surface layer to form centres of fines. Such sorted fine centres are seen on Victoria Island, where they extend downward in an almost cylindrical shape, sometimes resting on stones or gravel at depth.

Weathering has been suggested by some workers as an important process. F. Nansen (1922) stressed the importance of weathering, and suggested that depressions contained more moisture and that this accelerated the weathering, causing a concentration of fines in these areas. Once the fine centres had formed, freezing could push away the larger stones, as the fine centres were thought to freeze first. A similar view has been expressed by C. S. Elton (1927). He emphasized the importance of vertical differences in comminution. This would result in a fine layer underlying a coarser layer. The finer material could then be concentrated by differential upward heaving. Once the polygons had formed, the easier drainage through the coarser borders would make the process self-generating.

Contraction is another process that has been advocated to account for small desiccation polygons of the non-sorted variety. This process may also initiate sorted polygons if the border cracks are filled with stones by flooding. Such polygons are not necessarily associated with a periglacial climate. Contraction at low temperature will be considered later in connection with tundra polygons.

The convection theory was advocated by O. Nordenskjold (1909). He thought that sorted polygons could form as a result of aqueous convection currents, set up in the ground by the difference between the temperature at the surface and at the frost table. Although he later rejected the hypothesis, it was revived by A. R. Low (1925) and K. Gripp (1927). They suggested that convection resulting from the difference of density of water at 4°C and 0°C could cause currents. It is not generally thought that these would be capable of moving material.

A variant of this convection theory has recently been suggested by A. Journeaux and Ph. Bardey (1970). They suggest that the continuity of movement sets up convection cells with slow freezing. Particles gain water, lose density and, therefore, rise marginally with a central area of sinking. Large particles form the borders of sorted features because they cannot move once they reach the surface. With slow thawing, the movement is reversed, with central rising, outspreading and marginal sinking. A nest of stones forms as the large stones are left in the centre. The pattern of cells depends on viscosity, water volume, thermal conductivity, density, dilatation and mechanical characteristics of the material.

T. T. Paterson (1940) developed a theory based on a freezing mass that exerted pressure in an outward direction, perpendicular to the centre. This would result in the orientation of rocks parallel to the freezing surface. Rocks would stand vertically around the margin if the freezing centre were at the surface. This theory assumes wrongly that the ice pressure is exerted radially from a centre of fine material.

J. M. Anketell and S. Dzulynski (1968) describe transverse deformational patterns in unstable sediments when a high density layer overlies one of lower density. The lower layer may lose viscosity; liquefaction may then occur spontaneously followed by deformation. Experiments with plaster of paris and thixotropic clay produced rising material that formed a polygonal pattern, with narrow walls of plaster of paris forming linear features parallel to the boundary.

It will be apparent that many suggestions have been put forward. Most of them are based on some form of pressure generated by frost, caused by expansion of volume on freezing and subsequent reduction of pressure on melting. These changes can produce cryoturbation and sorting of material, where a heterogeneous mix of particles is available. Sorted circles, nets or polygons can result. From the experiments that have been described, it seems that the finer particles are moved farther and that these do not freeze so readily as the coarser centres. Thus the freezing centres do not consist of the fine but of the coarser particles. Alternating freeze and thaw does not require a permafrost climate to operate. The rapid temperature changes of high mountains provide suitable conditions in which freeze-thaw processes can operate. This helps to account for the common occurrence of sorted polygons in these areas.

5.2   *The formation of tundra or ice-wedge polygons*

The larger ice-wedge polygons, which do not involve much sorting of material, are formed by contraction caused by extreme cold. This view is now generally accepted. The mechanics of the process have been studied by A. H. Lachenbruch (1962) and described in Chapter 2. It is thought that the polygonal form results from the pattern of the contraction cracks. It is generally found that the diameters of the polygons are about twice or more the depth of the cracks. Lachenbruch has shown that polygonal cracking to form ice-wedge polygons is one example of a phenomenon that takes place in many different ways and in different media. The ice-wedge polygons form as a response to contraction under extreme cold. The thawing of the water that enters the cracks in summer causes troughs to form on the surface. When the tension near the surface is nearly as great as the tensile strength, the stress at a level 3–6 m lower may be compressive. This difference in stress is caused by a time lag in the penetration of the cold wave. The tension crack thus forms at the surface and is propagated downward.

J. R. Mackay (1973) has reported field observations of contraction cracking at Garry Island, N.W.T. The cracking occurred where tundra polygons were between 10–20 m in diameter, ranging from those with poorly defined troughs, through low-centred to high-centred polygons. During the period 1967–73, the percentage of ice wedges which cracked varied between 15 and 40, the majority of cracks taking place in February–March. Measurements showed that the cracks often extended down to between 3 and 5 m, and may be 5 mm wide at a depth of 3 m. The annual growth rate of the width of the ice wedge is an order of magnitude less than the width of the crack, because the cracks partially close before meltwater can seep into them to add to the ice wedge. Some are nearly entirely closed before the overlying snow can thaw and melt into them.

Cracks can form in a variety of patterns. One pattern is the orthogonal one, which occurs in heterogeneous material or plastic media. The cracks do not all form simultaneously and new ones tend to meet old ones at right-angles. The size of the polygon depends on the magnitude of the thermally-induced stress. Some of the orthogonal systems are random, but others are preferentially orientated. The latter normally occur in association with gradually receding water bodies, such as a shifting river channel or receding sea level. One set of cracks develops normal to the water body and the other at right-angles. This pattern is well exemplified on Foley Island in Foxe Basin. The cracks here are about 35 m apart and the more continuous ones are aligned parallel to the receding shoreline. The cracks are marked by double or triple ridges of limestone pebbles, the total width of the ridges being about 5 m (Plate VI).

The other type of tundra polygon is non-orthogonal and the cracks tend to meet at angles of 120°. These cracks form in homogeneous material subjected to uniform cooling. The cracks probably propagate laterally until they branch into two at a limiting velocity. The velocity again increases to a limiting value when renewed bifurcation takes place.

5.3   *The formation of the trough type of polygon*

This type of feature, described by Williams (1964) and others in the chalk areas of eastern England, may have a polygonal pattern on flat ground or become elongated on slightly sloping ground. There is an abrupt change of pattern between the upper flat surface and the slope beneath it. This change suggests that the features are not the result of gelifluction modifying a polygonal pattern on the slope as has been suggested by some workers. It appears that the features cannot be formed by ice wedges as these do not give rise to stripe forms, because even on quite steep slopes, ice wedges can form a polygonal pattern. The origin of the trough type of polygon is still not known.

5.4   *The formation of stripes*

Stripes are sometimes linked directly with polygonal features, developing from them as the gradient increases downslope. The sorted stripes are particularly likely to form as continuations of sorted polygons, for the same processes can operate, with modifications caused by gravity, on a slope. Like the sorted polygons, stripes are not confined to permafrost areas, and active stripe development is taking place on some of the hills of Britain. T. N. Caine (1963) has described currently developing stripes in the Lake District. He measured an upheaval of a fine stripe amounting to 4 cm after a cold spell, with a freezing period of 3–4 days, on Grasmoor at 760 m. Ground ice had formed during this period in thin parallel layers. The differential heave is thought to maintain the stripes. Hollingworth (1934) has also described striped scree on Blencathra in the Lake District. Some of the stripes had developed on old mine tips, and must, therefore, be less than 300 years old. They are actively forming at present. This has also been proved by R. Miller *et al.* (1954) by observations on the Tinto Hills. Experiments were carried out in this area by digging over the stripes to a depth of 30 cm. This destroyed the stripes, which had formed on a slope of about 20°. After two years, the stripes were beginning to reform. Differential heaving and better drainage in the coarser elements were thought to cause the sorting of particles and to account for the stripes.

It should be emphasized that not all stripes are connected with periglacial activity, just as in the case of some forms of unsorted polygonal markings. Striped slopes, for instance, occur in some semi-arid and sub-tropical areas, and their possible origins will not be discussed here. Stripes that do not involve frost sorting but nevertheless are to be found in areas of frost climate must not, however, be neglected. Needle ice (p. 101) is one possible agent in creating micro-patterns of stripes. J. R. Mackay and W. H. Matthews (1974) discuss the formation of such stripes in British Columbia and New Zealand. The stripes form as a result of melting of the needle ice on sunny mornings, the orientation of the stripe being parallel to the sun's shadow when collapse took place. The striping appears to be caused by differential melting of the soil cap and needles, combined with lateral movement of some of the soil particles.

The Antarctic islands of Crozet and Kerguelen have well-developed soil stripes (P. Bellair, 1969). The slopes on which soil stripes are well developed are very wet, partly because of the impermeability of the underlying bedrock. Permafrost is unlikely to exist in this area, where the mean annual temperature exceeds 0°C. The patterned ground is probably caused by alternate wetting and drying. Although the surface is generally very wet, winds cause rapid drying during brief intervals, leaving a lag gravel owing to deflation of the dried surface. The stripes do not appear to be connected with polygons or ground freezing in this area.

6 Conclusions

Patterned ground is one of the most conspicuous elements of many periglacial landscapes. The patterns can be classified according to their form into circles, nets and polygons, each of which can be sorted or unsorted in character. These shapes are restricted to fairly flat ground. On steeper slopes of moderate gradient, steps and stripes occur. Polygons are the most widespread type of pattern and these can be subdivided into two main types. The smaller type is most characteristic of high mountains in middle latitudes that are subject to freeze-thaw but not necessarily to permafrost. The larger type, called tundra or ice-wedge polygons, only form in areas of active permafrost and low mean annual temperatures of at least −6°C.

Many theories have been put forward to explain the first type. Some form of frost action causing cryoturbation and sorting of mixed material is involved in their formation, although the exact process is not clearly established. Observations and experiments have been made to assess the validity of the suggested processes. The nature of the material has been found to play an important part. The tundra polygons are generally agreed to form by contraction cracking caused by intense cold in winter. The cracks fill with ice; later, they may become fossilized by being filled with sediment when the ice melts. The existence of fossil ice wedges and their associated surface polygonal pattern gives valuable evidence on palaeoclimatic conditions. The stripes are mostly formed by similar processes to the smaller, sorted polygons, where these processes act on slopes up to 30°.

7 References

ANKETELL, J. M. and S. DZULYNSKI, S. (1968), 'Transverse deformational patterns in unstable sediments', *Annls Soc. géol. Pologne* **38**, 411–16

BALL, D. F. and GOODIER, R. (1970), 'Morphology and distribution of features resulting from frost action in Snowdonia', *Field Stud.* **3**, 193–218

BELLAIR, P. (1969) 'Soil stripes and polygonal ground in the Sub-antarctic islands of Crozet and Kerguelen' in *The periglacial environment* (ed. T. L. PÉWÉ), Montreal, 217–22

BENEDICT, J. B. (1969), 'Microfabric of patterned ground', *Arct. alp. Res.* **1**, 1 45–8

BLACK, R. F. (1954), 'Permafrost—a review', *Bull. geol. Soc. Am.* **65**, 839–55

(1963), 'Les coins de glace et la gel permanent dans le Nord de l'Alaska', *Annls Géogr.* **72**, 257–71

BLACK, R. F. and BERG, T. E. (1963), 'Glacier fluctuations recorded by patterned ground, Victoria Land', *Antarct. Geol.* (S.C.A.R. Proc., 1963), **3**, 1, 107–22

BUNTING, B. T. and JACKSON, R. H. (1970), 'Studies of patterned ground on south-west Devon Island, N.W.T. A method of collecting undisturbed megasamples from dry arctic polygons. Observations on some physical properties of the material in a sorted polygon', *Geogr. Annlr* **52**A, 194–208

CAILLEUX, A. and ROUSSET, C. (1968) 'Presence de réseaux polygonaux de fentes en coin en Basse-Provence occidentale et leur signification paleoclimatique', *C. r. Séance Acad. Sci.* **266**D, 669–71

CAINE, T. N. (1963), 'The origin of sorted stripes in the Lake District, northern England', *Geogr. Annlr* **45**, 172–9

CLARK, G. M. (1968), 'Sorted patterned ground: new Appalachian localities south of the glacial border', *Science* **161**, 355–6

CORBEL, (1954), 'Les sols polygonaux: observations, expériences, genèse', *Revue Géomorph. dyn.* **5**, 49–68

CORTE, A. E. (1962), 'Horizontal sorting. The frost behaviour of soils: laboratory and field data for a new concept', *U.S. Army Cold Reg. Res. Engng Lab., Res. Rep.* **85**, 2, 20 pp.

(1963), 'Relationship between four ground patterns, structure of the active layer, and type and distribution of ice in the permafrost', *Biul. Peryglac.* **12**, 7–90

(1966), 'Particle sorting by repeated freezing and thawing', *Biul. Peryglac.* **15**, 175–240

DIMBLEBY, G. W. (1952), 'Pleistocene ice wedges in north-east Yorkshire', *J. Soil. Sci.* **3**, 1–19

DIONNE, J.-C. (1969), 'Nouvelles observations de fentes de gels fossilés sur la côte sud du Saint-Laurent', *Rev. Géogr. Montréal* **23**, 307–16

DREW, J. V. and TEDROW, J. C. F. (1962), 'Arctic soil classification and patterned ground', *Arctic* **15**, 109–16

DYBECK, M. W. (1957), 'An investigation into soil polygons in central Iceland', *J. Glaciol.* **3**, 143–6

ELTON, C. S. (1927), 'The nature and origin of soil polygons in Spitsbergen', *Q. J. geol. Soc. Lond.* **83**, 163–94

FITZPATRICK, E. A. (1958), 'An introduction to the periglacial geomorphology of Scotland', *Scott. geogr. Mag.* **74**, 28–36

FRIEDMAN, J. D., JOHANSSON, C. E., OSKARSSON, N., SVENSSON, H., THORARINSSON, S. and WILLIAMS, R. S. (1971), 'Observations on Icelandic polygon surfaces and palsa areas. Photographic interpretation and field studies', *Geogr. Annlr* **53**A, 115–45

GRIPP, K. (1927), 'Beiträge zur Geologie von Spitsbergen', *Abh. Geb. Naturw.,*
Hamburg **21**, 1–38

HAMBERG, A. (1915), 'Zur Kenntnis der Vorgange im Erdböden beim Gefrieren
und Auftauen so wie Bemerkungen über die erste Kristallisation des Eises im
Wasser', *Geol. För. Stockh. Förh.* **37**, 583–619

HENDERSON, E. P. (1968), 'Patterned ground in south-east Newfoundland', *Can. J.
Earth Sci.* **5**, 1443–53

HÖGBOM, B. (1910), 'Einige Illustrationes zu den geologischen Wirkungen des
Frostes auf Spitsbergen', *Bull. geol. Instn Univ. Upsala* **9**, 41–59
   (1914), 'Über die geologische Bedeutung des Frostes', *Bull. geol. Instn Univ.
Upsala* **12**, 257–389

HOLLINGWORTH, S. E. (1934), 'Some solifluction phenomena in the northern part of
the Lake District', *Proc. Geol. Ass.* **45**, 167–88

JOURNEAUX, A. and BARDEY, P. (1970), 'Réflexions sur les sols polygonaux, les sols
striés et les nids de pierre', *Acta Geogr. Lodziensia* **24**, 249–58

KÄLLENDER, H. (1967), 'Polygonal ground and solifluction features. Photographic
interpretation and field studies in north Scandinavia', *Lund Stud. Geogr.* **40**A,
24–40

KING, R. B. (1971), 'Boulder polygons and stripes in the Cairngorm Mountains,
Scotland', *J. Glaciol.* **10**, 375–86

LACHENBRUCH, A. H. (1962), 'Mechanics of thermal contraction cracks and
ice-wedge polygons in permafrost', *Geol. Soc. Am. Spec. Pap.* **70**, 69 pp.

LEFFINGWELL, E. K. (1915), 'Ground ice wedges', *J. Geol.* **23**, 635–54

LOW, A. R. (1925), 'Instability of viscous fluid motion', *Nature, Lond.* **115**, 299–300

MAACK, A. (1967), 'Polygonal ground and solifluction features. Photographic
interpretation and field studies in north Scandinavia', *Lund Stud. Geogr.* **40**A,
41–57

MACKAY, J. R. (1973) 'Winter cracking (1967-73) of ice-wedges, Garry Island,
N.W.T.', *Geol. Surv. Can. Pap.* **73–1**B, 161–3

MACKAY, J. R. and MATHEWS, W. H. (1974), 'Needle ice striped ground', *Arct. alp.
Res.* **6**, 79–84

MARR, J. W. (1969), 'Cyclical change in a patterned-ground ecosystem, Thule,
Greenland' in *The periglacial environment* (ed. T. L. PÉWÉ), Montreal, 177–201

MILLER, R., COMMON, R. and GALLOWAY, R. W. (1954), 'Stone stripes and other
surface features of Tinto Hill', *Geogrl J.* **120**, 216–19

MORGAN, A. V. (1971), 'Polygonal patterned ground of late Weichselian age in the
area north and west of Wolverhampton, England', *Geogr. Annlr* **53**A, 146–56
   (1972), 'Late Wisconsinan ice-wedge polygons near Kitchener, Ontario,
Canada', *Can. J. Earth Sci.* **9**, 607-17

NANSEN, F. (1922), *Spitsbergen* (Leipzig, 3rd Ed.), 327 pp.

NORDENSKJOLD, O. (1909), *Die Polarwelt* (Leipzig and Berlin), 220 pp.

OHRNGREN, S. (1967), 'Polygonal ground and solifluction studies in north
Scandinavia', *Lund. Stud. Geogr.* **40**A, 58–67

PATERSON, T. T. (1940), 'The effects of frost action and solifluction around Baffin
Bay and in the Cambridge district', *Q.J. geol. Soc. Lond.* **96**, 99–130

PÉWÉ, T. L. (1963), 'Ice wedges in Alaska—classification, distribution and climatic significance', *Geol. Soc. Am. Spec. Pap.* **76**, 129

PHILBERTH, K. (1964), 'Recherches sur les sols polygonaux et striés,' *Biul. Peryglac.* **13**, 99–198

PISSART, A. (1964), 'Vitesse des mouvements du sol au Chambeyron (Basses Alpes)', *Biul. Peryglac.* **14**, 303–10

(1968), 'Les polygones de fente de gel de l'Île Prince Patrick (Arctique Canadien—76°lat. N.)', *Biul. Peryglac.* **17**, 171–80

RAPP, A. and CLARK, G. M. (1971), 'Large non-sorted polygons in Padjelanta National Park, Swedish Lapland', *Geogr. Annlr* **53**A, 71–85

RAPP, A. and RUDBERG, S. (1964), 'Studies on periglacial phenomena in Scandinavia', *Biul. Peryglac.* **14**, 75–90

SHOTTON, F. W. (1960), 'Large-scale patterned ground in the valley of the Worcestershire Avon', *Geol. Mag.* **97**, 404–8

SVENSSON, H. (1969), 'Open fissures in a polygonal net on the Norwegian Arctic coast', *Biul. Peryglac.* **19**, 389–98

TABER, S. (1930), 'The mechanics of frost heaving', *J. Geol.* **38**, 303–17

(1943), 'Perennially frozen ground in Alaska: its origin and history', *Bull. geol. Soc. Am.* **54**, 1433–548

THORARINSSON, S. (1964), 'Additional notes on patterned ground in Iceland with a particular reference to ice-wedge polygons,' *Biul. Peryglac.* **14**, 327–36

TUFNELL, L. (1969), 'The range of periglacial phenomena in northern England', *Biul. Peryglac.* **19**, 291–323

WASHBURN, A. L. (1956), 'Classification of patterned ground and review of suggested origins', *Bull. geol. Soc. Am.* **67**, 823–66

(1969), 'Patterned ground in the Mesters Vig district, north-east Greenland', *Biul. Peryglac.* **18**, 259–330

(1970), 'An approach to a genetic classification of patterned ground', *Acta Geogr. Lodziensia* **24**, 437–46

WATT, A. S., PERRIN, R. M. S. and WEST, R. G. (1966), 'Patterned ground in Breckland: structure and composition', *J. Ecol.* **54**, 239–58

WILLIAMS, P. J. (1961), 'Climatic factors controlling frozen ground phenomena,' *Geogr. Annlr* **43**, 339–47

WILLIAMS, R. B. G. (1964), 'Fossil patterned ground in eastern England', *Biul. Peryglac.* **14**, 337–49

4

Periglacial mass movements and slope deposits

As noon passed, the soil in all the hollows or small watercourses became semifluid and very uncomfortable to walk on or sink into. The entire slope, in consequence of the thaw, had become a fluid moving chute of debris for at least one foot in depth. (SIR EDWARD BELCHER on Buckingham Island, 77°N., quoted by J. Geikie, 1874)

Certain features of the periglacial environment are particularly favourable to mass movement of weathered debris (sometimes loosely described as 'soil' or 'earth', though the existence of a soil profile is not normally implied). Alternations of freezing and thawing will disturb the debris in various ways; meltwater from snow, ice or ground ice will soak the debris and facilitate its movement; frozen ground at depth prevents downward percolation of moisture in thawed surface layers; and the vegetative cover may be too restricted to prevent mass movements even on relatively gentle slopes. J. G. Andersson's observations (1906) on Bear Island led him to propose the term 'solifluction' for 'the slow flowing from higher to lower ground of masses of waste saturated with water' (p. 95). Andersson recognized, however, that solifluction thus defined was not restricted to areas of any one type of climate; he maintained nevertheless that a sub-arctic climate (Andersson—'subglacial') provided optimum conditions, and that in such regions, solifluction was the chief agent of denudation: 'here, the removal of waste runs on at a rate that may be unsurpassed in other parts of the earth's surface ...' (p. 112).

Since the literal meaning of 'solifluction' is 'soil flow', it covers a number of different processes of mass movement occurring under a variety of climatic conditions, and also varying rates of movement, from slow downhill creep to relatively rapidly moving mud flows. Opinion is divided as to whether 'solifluction' should be used in such a wide sense. A. Cailleux and J. Tricart (1950) have used the term to include both slow creep and rapid mud slides under climatic conditions varying from periglacial to moist temperate. P. D. Baird and W. V. Lewis (1957) describe as 'solifluction' certain mass movements in the Cairngorms resulting from the heavy rainstorms of the summer of 1956. A. Rapp (1962), however, distinguishes sharply between rapid sporadic movements such as mud flows and debris slides, the slow creep of coarse talus on steep slopes, and the equally slow creep of finer grained material on less steep slopes. Only the latter of these movements he is willing to designate as solifluction. J. Dylik (1951) proposed the term 'congelifluction' to describe earth flow in the presence of permafrost, but excluding frost creep, sheetwash and rapid mud flows, while K. Bryan (1946) had earlier suggested 'congeliturbation' to comprise all mass movements under periglacial conditions, including solifluction.

The term 'gelifluction' was used by H. Baulig in 1956 (a,b):

'en régime cryergique, l'ablation des débris se fait surtout par solifluxion—on dirait peut-être plus précisément par *géli(soli)fluxion*. Celle-ci consiste dans le déplacement lent de matériaux dégelés, saturés d'eau, glissant sur un sous-sol encore gelé ...' (Baulig, 1956b, pp. 50–51).

The advantage of using this term in place of solifluction is that a periglacial régime is definitely implied. Although many workers still use the term solifluction, there is a growing tendency to use gelifluction (preferred to Dylik's congelifluction because it is shorter), with its more precise connotation, to denote the flow of thawed material over frozen sub-surface layers. A. L. Washburn (1973) adopts this term to cover processes of downslope movement, excluding frost creep, in the periglacial zone. L.-E. Hamelin and F. A. Cook (1967) retain 'solifluction' in the English section of their glossary, but adopt 'gélifluxion' in the French. For processes of ground movement in which heaving and sorting predominate, the term 'cryoturbation' is preferable.

It is unfortunate that gelifluction has been used both to denote the combined effect of periglacial slope processes in general and for a particular process (the flow of thawed debris) excluding frost creep. As will be discussed in the next section, there are often practical problems of differentiating processes of downslope movement in the field and also in the laboratory. Although gelifluction (*sensu stricto*) may be more rapid than frost creep, the two processes may often operate simultaneously at certain times of the year. A. Jahn (1967) supports R. S. Sigafoos and D. M. Hopkins (1952) in their contention that solifluction (gelifluction *sensu lato*) may take place by simultaneous creep and flow.

In this chapter, the term gelifluction will be used *sensu stricto* to signify the flow of water-soaked debris over permanently or seasonally frozen sub-surface layers, akin to mud flow, and the term frost creep employed to describe movements induced by alternate freezing and thawing of debris resting on a slope. In both, gravity is the motive force, and the periglacial environment provides unstable conditions in the slope debris encouraging its downhill movement. The discussion will also be confined to what C. Troll (1944) referred to as 'macro-solifluction', excluding small-scale forms of movement such as occur in stone sorting within polygons and which Troll describes as 'micro-solifluction'.

1 Gelifluction

Permanently or seasonally frozen sub-surface layers prevent downward percolation of moisture. The upper layer affected by seasonal thawing (the 'active layer') becomes soaked with water from melting snow, from melting of any temporary segregations of ice within it, or from rainfall. Washburn (1947) in part of the Canadian Arctic found that the melting of ground ice was the principal source of moisture, since the climate is semi-arid, most rainfall comes in autumn (the active layer is already water-saturated long before this in the spring) and most slopes have lost their snow cover by sublimation or wind action before any melting occurs. S. Taber (1943) also emphasized the importance of ground ice in concentrating excessive amounts of water in the active layer, a process unknown to Andersson. Taber notes how some frozen silts in Alaska contain over 80 per cent of ice by volume* and would convert to liquid mud on thawing. In the case of massive ground ice observed in Tuktoyaktuk Peninsula, N.W.T., Canada, J. R. Mackay (1972) notes that melting of the ice-rich sediments would produce about 95 per cent water and only 5 per cent sediment by volume. In other regions, the contribution of snow-melt seems to be of greater importance. This is borne out by P. J. Williams' study (1957) of gelifluction below snow-patches in Rondane, Norway. These appeared to melt substantially from the underside, soaking the mantle of debris beneath and downslope from the snow-patch. The debris in this case was not significantly frost-susceptible (Chapter 2, p. 34) so that melt from ground ice was negligible. However, studying gelifluction in frost-susceptible materials in the Dovrefjell area, Williams concluded that even in these, ice-layer melting did not represent a water increase that could by itself have caused flow to occur. It is likely that snow-melt is a more important source of water in some areas, and that in others, ground ice may be a major contributor. Differences of climate and of frost-susceptibility of the materials must both be considered.

The effect of excess water in the active layer is to reduce its shear strength. The general explanation of the process as one of lubrication of soil particles by water is incomplete and misleading: Williams (1959) observes, for instance, that water acting on quartz grains is not a lubricant. The shear strength of a material depends on internal friction and cohesion. The former depends in turn on the pressure at the contacts between individual grains and this varies with the pore-water pressure. The magnitude of cohesion in saturated debris depends on water content and therefore on the ratio of voids to solid. Significant gelifluction probably only occurs, according to Washburn (1973), at moisture values corresponding to or exceeding the Atterberg liquid limit, when soils have very little, if any, shear strength. Soaking of debris with water thus reduces both internal friction and cohesion (it will be noted later that frost-heave has a similar effect), and it also increases the weight of the material. Snow patches resting on the active layer may also contribute to instability by their weight (see Chapter 5), increasing the shear stress; on the other hand, when the snow is thickest, the ground may well be frozen up to the surface and easily capable of bearing the load, and at other times of the year, residual frozen ground beneath a surviving snow patch may serve to spread the load.

Vegetation acts as a most important restraining factor and there is also an important interaction, for gelifluction in turn affects the vegetation. Turf and even moss may be able to hold the semi-fluid layer in place until the load becomes too great. S. R.

---

* We are grateful to Dr J. Hutchinson for pointing out that Taber (1943) writes '80 per cent of ice by weight' on p. 1457, but on p. 1495 quotes the same percentage by volume. By comparison with other data (both from Taber and from others) the figure of 80 per cent should refer to volume.

Capps' descriptions (1919) of the phenomena in Alaska have never been improved on. In an area of permafrost, where seasonal thawing takes place to depths of a metre or so and where melting snow supplies water, movements of debris vary in rate from slow to rapid according to ground slope and the strength of the surface mat of grasses and mosses. At times on steep slopes, the turf ruptures suddenly, liberating rapid mud-flows and leaving scars. The slower and sometimes imperceptible flow movements are more widespread, however, and were considered by Capps to be quantitatively more important. A. J. Broscoe and S. Thomson (1972) recorded an example of very rapid mudflow in the St. Elias Mountains of Alaska in 1967. Following 50 mm of rain in 36 hours, the flow built up moving arcuate fronts 2–3 m high which advanced, with increasing radii, in pulses owing to the creation of temporary dams. Boulders 4 m in size were borne along in the slurry. Such events are not, of course, peculiar to the periglacial environment, and are common in alpine regions of steep slopes, high precipi-tation and snow-melt. They represent an extreme case of water saturation of superficial debris and its downslope movement. The distinction between mudflow and gelifluction is arbitrary; it is best to reserve the term gelifluction for superficial movements over a frozen sub-surface.

2 Frost creep

During freezing of a frost-susceptible material, ice crystals grow normal to the cooling surface and displace particles in this direction. On thaw, the particles re-settle approxi-mately in a direction controlled by gravity. Thus if the cooling surface is inclined, the displaced particles will always re-settle slightly downhill from their original positions. The process has long been known. C. Davison in 1889 successfully reproduced such creep in tilted soil-filled boxes exposed to repeated freezing and thawing. Tricart reported similar but more closely controlled experiments in 1954. The amount of po-tential creep can be readily computed: for a $10°$ slope, frost-heaving of 10 cm will give a theoretical maximum downslope movement of $1·76$ cm, though exactly vertical con-solidation after freezing is unlikely. The amount of movement will also decrease from the surface downward (Williams, 1957) giving a curved velocity profile; it is this which is mainly responsible for the turning of stones and boulders to lie parallel to the surface.

In his original experiments reported in 1889, Davison noted a tendency for the par-ticles displaced by frost-heave to settle back against the slope during contraction, rather than purely vertically. The term 'retrograde movement' is now used to denote this phenomenon; true frost creep is therefore given by potential frost creep minus the retro-grade movement. The process of retrograde movement is not fully understood: Wash-burn (1967) suggests that capillary pressure, which must be orientated normal to the slope in a uniform material, will tend to contract the material along the normal during drying and settling, and so tend to act against gravity, often being many times stronger than the latter. This effect will occur on saturated slopes as well as on drier slopes so long as moisture is being lost by capillary movement to an evaporating surface.

Frost-heave also reduces the shear strength of a material. In the first place, ice layer formation will disrupt frost-susceptible materials, forming discontinuities at the ice surfaces which will be significant planes of weakness when the ice melts. Secondly, frost-heave increases the void ratio and therefore permeability, and in turn reduces cohesion between the particles. These conditions persist during thaw, and allow the material to soak up more water if it is not already saturated.

Williams (1957, 1959) has studied the part played by frost-heave in downslope movement in Norway. Below a snow patch, movement was measured over periods of several weeks in successive years on a slope averaging 19°. No permafrost was observed. The grain-size composition of the debris (see the continuous line for one sample on Fig. 2.5) indicated a high frost-susceptibility, conformed by the existence of free ice layers amounting to one-fifth of a 15 cm soil column. In this case, there was strong evidence

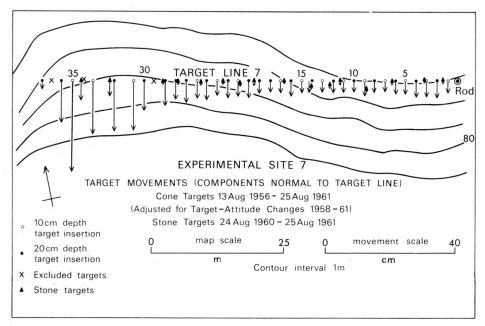

Fig. 4.1
Downslope movements recorded along a target line in the Mesters Vig area of north-east Greenland (A. L. Washburn, *Meddr Grønland*, 1967)

that frost-heave was the principal process of downslope movement. There was no significant change in water content of the debris as between spring and late summer, and little increase in soil weight during the thaw period. Unexpectedly low pore-water pressures were recorded, resulting from increases in permeability consequent on frost-heaving. Rates of downslope movement were small (see p. 103), typical of slow creep caused by frost-heave.

An extensive investigation of downslope movements by both gelifluction and frost creep has been carried out by Washburn (1967) in the Mesters Vig district of north-east Greenland. Observations at many sites comprised several thousand theodolite readings on cone targets inserted to depths of 10 and 20 cm, on gradients ranging from 2 to 25°. Fig. 4.1 shows one of the experimental sites and the results obtained. Table

4.1 summarizes the annual movements at some sites, and the significant relationships between the availability of moisture at a site, the local gradient and the downslope movement. The role of moisture is clearly fundamental, playing a more important part than gradient and, also, vegetation. In this district, paradoxically, the most rapid movements were in the best vegetated sector. Washburn finds it possible to establish the approximate relative contributions of gelifluction and frost creep throughout the year. In winter, frost-heaving normal to the slope is dominant; in the transition to summer, gelifluction gradually takes over and particles move mainly parallel to the surface. September is a month when both gelifluction and frost creep occur, while later in the year frost-heave again becomes dominant and rates of movement decline.

In the Colorado Front Range, J. B. Benedict (1970) has documented downslope movements for a high-altitude, lower latitude region. Potential frost creep here began in November (Fig. 4.2) and continued till May, movement on Lobe 45, for instance, ranging from 1 to 22·5 mm (greater than is theoretically possible under existing heave

Table 4.1 Downslope movements at some sites in the Mesters Vig district, north-east Greenland (A. L. Washburn, 1967)

Site	Mean gradient		Movement, cm/yr	
	Dry*	Wet*	Dry*	Wet*
6	2·5°	2·5°	—	1·0
7	12·5°	10·5°	0·9	3·4
8	12·5°	11·5°	2·9	3·7
15	—	3·0°	—	3·1
16	25·0°	—	1·1	—
17	—	12·0°	—	12·4

*Dry and wet refer to variations in ground moisture along the target lines at each site.

and slope conditions, but it is not clear why this should be so). Gelifluction began in May during the spring thaw, and accounted for up to 22 mm of movement. By mid-July, all stakes had started to move upslope from 9 to 31 mm (the 'retrograde movement'), stopping only with the next winter freezing (Fig. 4.2B). Variations in rates of movement were found to be closely related to variations in soil moisture, as Washburn found, and to a lesser extent to gradient. C. Harris (1972) also found that moisture content was more important than slope in a study of turf-banked lobes in northern Norway.

The formation of needle-ice (piprake, mush frost, or efflorescent ice) on slopes gives rise to downhill creep very similar to that caused by frost-heave, but it is usually a nightly rather than a seasonal phenomenon. Clusters of ice needles growing at the surface may be capable of lifting pebbles as well as finer debris several centimetres; when the crystals melt or break, the particles drop back to a lower level on the slope. Troll (1944) terms it 'needle-ice solifluction'. Evidence of it having occurred on the ground usually takes the form of disrupted vegetation and, especially on coarser mineral soils, a 'fluffy' appearance to the surface. C. Schubert (1973) describes a form of 'striated' surface in the Andes, caused by rows of needle-ice forming parallel to the prevailing

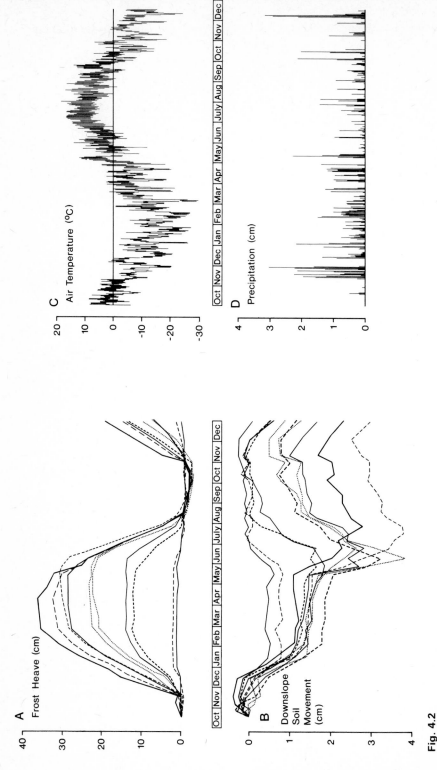

Fig. 4.2

Frost heaving and downslope movement for Lobe 45, Colorado Front Range (J. B. Benedict, *Arctic and Alpine Research*, 1970)

A Heaving and settling of the ground surface alongside movement stakes, October 1965–December 1966.

B Displacement of 25 cm stakes, parallel to the ground surface.

C Daily air temperature fluctuations at nearby recorder.

D Daily precipitation at nearby recorder

wind. The needles attained heights of 4–5 cm, pushing up small particles which, on thawing of the crystals which were slightly inclined down-wind, toppled off to form faint lines of debris.

The relative importance in downslope movement of 'flow' and 'creep' varies according to local circumstances. Creep will only be significant in frost-susceptible materials and will be much more important in areas with many freeze-thaw cycles. It is probably relatively unimportant in a high polar environment; in contrast, Benedict finds it the dominant process at high levels in the Colorado Rockies. In practice, the two processes often overlap to such an extent, particularly at certain times of the year, that their separation is difficult if not impossible. Among the factors that must be considered in assessing the relative importance of creep and flow are the grain-size composition of the material, the availability of water, the depth of frost penetration and the number of freeze-thaw cycles, and the competence of the vegetative cover.

The combined operation of the two processes has been studied in laboratory experiments by A. Higashi and A. E. Corte (1971), using a tilted box filled with a very frost-susceptible silty clay continually supplied with water from below, slope angles of $3°$, $7\frac{1}{2}°$ and $15°$, and a constant series of freeze-thaw cycles. During these freeze-thaw cycles, frost creep was dominant, mostly at and close to the surface. Ice needles formed during freezing frequently bent downslope under their own weight, adding to the amount of theoretical frost creep. At the end of the experiment using a $3°$ slope, the soil was allowed to thaw out completely for one week. This produced much more movement at depth in the soil, attributed to gelifluction in the presence of excess water.

3 Rates of movement

There are various methods of measuring gelifluction and frost-creep movements. Painted markings or stakes can record surface displacements (for example, Rapp, 1962; Washburn, 1967; Benedict, 1970). Velocity profiles may be obtained by sinking an iron pipe, filling this with small plastic or wood cylinders one above the other, withdrawing the pipe, and then at some future date carefully exhuming the small cylinders and recording their changed positions (S. Rudberg, 1962) (Fig. 4.3). Alternatively, lead cable or polythene tubing cut into short lengths can be packed into a borehole and exhumed in the same way, or electrical strain gauges can measure the amount of bending of metal strips (Williams, 1957).

Rates of surface movement vary with local conditions as expected. The following are some examples of movements occurring partly or wholly by creep.

A. L. Washburn (1947), Canadian Arctic. Maximum 3·8 cm in one month.

P. J. Williams (1959), Dovrefjell. Slope 19°. Maximum 25 cm in 3 weeks. One movement of 10 cm in 24 hours recorded.

A. Jahn (1960), Spitsbergen. Slope 7–15°. Maximum 12 cm a year.

J. Smith (1960), South Georgia. Slope 21°. Maximum 5 cm a year. Surface pebbles moved up to 70 cm a year.

A. Rapp (1962), Sweden. Slope 15–25°. Maximum 30 cm a year.

S. Rudberg (1962), Sweden. Slope 15°. Maximum 7·5 cm a year (in this case, the bulging front of a gelifluction terrace).

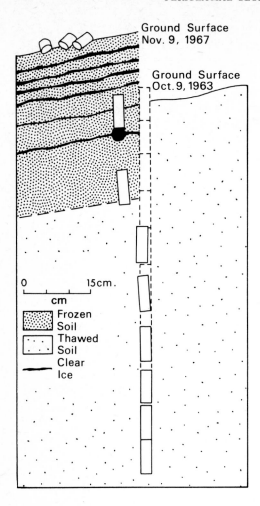

Ground Surface
Nov. 9, 1967

Ground Surface
Oct. 9, 1963

0   15cm.

cm

Frozen
Soil

Thawed
Soil

Clear
Ice

Fig. 4.3
Velocity profile, Lobe 45, Colorado Front
Range (J. B. Benedict, *Arctic and Alpine
Research*, 1970). The column of cylinders
was placed in a vertical hole on 9 October
1963 and exhumed, while partially frozen,
on 9 November 1967. Maximum movement
of 7·7 cm/year occurred at the surface
where cylinders heaved out of the soil were
affected by diurnal freeze-thaw. Downslope
movement persisted to a depth of about
50 cm.

J. B. Benedict (1970), Colorado Front Range. Maximum 4·3 cm a year.

C. Harris (1972), Okstindan, north Norway. Slope 5–17°. Maximum 6 cm a year
(see also Table 4.1 for some of the data assembled by Washburn, 1967)

All observers find that movements are spasmodic (short-term movements must not be
extrapolated for longer periods) and are in agreement that rates of movement are less
than was previously thought. As Rapp points out, B. Högbom's (1914) suggestion of

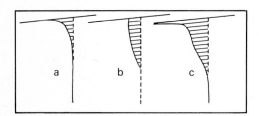

Fig. 4.4
Graphical addition of (a) a typical vertical
velocity profile resulting from frost creep and
surficial freeze-thaw and (b) a typical profile
resulting from gelifluction unimpeded by
any restraining turf layer, produces (c) a
velocity profile similar to that shown by Fig.
4.3 (J. B. Benedict, *Arctic and Alpine Research*, 1970).

one or more metres a year for the upper limit of movement by *creep* seems likely to be a ten-fold overestimation. On the other hand, local and sporadic *flow* movements can be much more rapid—several metres in a few days under optimum conditions according to A. Jahn (1967, p. 218).

Sub-surface records of movement show the expected and relatively constant diminution of velocity with depth (Figs. 4.3 and 4.4). Rudberg (1962) gives the following velocity profile for a site at Kärkevagge in Sweden:

Surface	5 cm/year
−10 cm	3·8 cm/year
−20 cm	2·8 cm/year
−30 cm	1·7 cm/year
−40 cm	0·8 cm/year
−50 cm	0·3 cm/year

4 The climatic environment

Gelifluction and frost creep may occur in areas of ground frozen permanently, seasonally, or diurnally (Troll, 1944). Permafrost is not essential. The requirements in areas lacking permafrost are fairly deep and rapid frost penetration followed by thawing from the surface downward. In the case of gelifluction in areas of seasonally frozen ground, summers must be relatively cool, as otherwise heat stored in the ground from the previous summer will promote thawing from below the level of winter frost penetration; this will be adverse to the preservation of a frozen layer in spring. Gelifluction and frost creep on a large scale are therefore more characteristic of areas not experiencing warm summers and where the mean annual air temperature is not higher than 1°C (Williams, 1961). They appear to need a more severe climate than certain other periglacial phenomena such as frost hummocks.

At the present day, the lower limits of gelifluction and frost creep in Britain appear to be at about 970 m in the Cairngorms, and possibly as low as 370 m in the Shetlands where windy conditions prevent more than a scanty vegetative cover (R. W. Galloway, 1961). In central Europe, the lower limit is about 2000 m in the central Alps and 1500 m in the Riesengebirge (J. Büdel, 1944). In the Pleistocene, gelifluction and frost creep were probably ubiquitous in central Europe, for their deposits are widespread and well preserved; Büdel (1937) claimed that the processes were then active on slopes as low as 2° and that debris had moved as much as 2 km. In Britain, gelifluction was active down to sea level in the last glaciation, even in southern Britain, and in the Late-glacial (Zone III), gelifluction deposits are found as low as 230 m on Bodmin Moor (A. P. Conolly *et al.*, 1950).

5 Periglacial slope deposits

Numerous terms have in the past been used to designate such deposits. In 1839, for instance, H. T. de la Beche put forward the term 'head', still familiar to British geologists; in 1887, Clement Reid explained the origin of 'coombe rock' as the natural gelifluction deposit of chalk country. Another term used from time to time in the past is 'trail' (H. G. Dines *et al.*, 1940). In 1946, Kirk Bryan proposed the name 'congeliturbate' and advocated the abandonment of local terms, but since congeliturbate as he defined it is the product of all movements resulting from frost-heave and the flow of water-saturated debris, it can include deposits other than those affected by downslope mass movements.

Gelifluction deposits are closely related to local rock type since it is rare for material to have travelled farther than 1 or 2 km by the processes described. Thus, unlike glacial

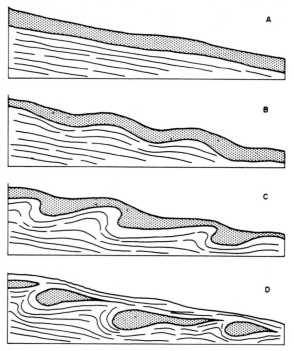

Fig. 4.5
The development of plications in slope deposits by downslope movement (A. Jahn, *Biul. Peryglac.*, 1956)

drift, erratics are rare (unless the source of the material is drift itself). Angularity of the fragments is characteristic since they have not travelled far, though some reduction in angularity with increasing distance from the source has been occasionally noted (for instance, T. T. Paterson, 1951, p. 11). The thickness of the deposits is extremely variable. Depressions and valley bottoms may collect considerable infills, sometimes several tens of metres thick. The longer axes of stones in the deposit lie in the direction of movement (compare glacial drift) until the point is reached where movement is arrested, as at the edge of a lobe or terrace. Then the blocks are suddenly turned at right-angles to the previous movement (G. Lundqvist, 1949). Reorientation at the border is facilitated by the fact that blocks never reach the front of the moving mass at exactly 90° to it. Obstacles within the moving mass also cause local reorientation of stones. Flow orientation is less persistent and regular in the lower layers of such deposits, unless the latter are derived from glacial drift, when orientations in the lower layers may express the last direction of ice motion. Signs of bedding are usually rare: head deposits are typically massive and unsorted. The origin of occasional indistinct bedding is not clear, but it may represent accumulation in different episodes and may be related to changing climatic conditions.

Contorted structures or 'involutions' are often typical of those deposits where bedding can be distinguished. Involutions associated with ground-ice formation were considered in Chapter 2. Downslope movement can itself produce deformation or drag structures owing to differential lateral movements: K. Byran (1946) distinguished these as 'plications'. A. Jahn (1956) shows stages in the development of such forms, the final stage being characterized by roll-like or cylindrical masses (Fig. 4.5).

Several problems arise in distinguishing head deposits from glacial drift. The problem is particularly intractable if the deposit has itself originated from drift. But apart from this, periglacial slope deposits may sometimes be distinguishable by the downslope orientation of stones, the more angular nature of these stones, the absence of erratics, and the generally looser texture.

An example of typical Quaternary head deposits from south-west England is given by D. N. Mottershead (1971). Coastal sections expose thicknesses of at least 33 m, where the deposits have buried a previous coastal slope and cliff-line and spread over old marine beaches. At their upper limit, the deposits merge with blockfield debris above which rise tors (see Chapter 6). Mottershead's analysis of numerous samples shows that the head is coarse (mean particle size of 18 samples $= -1\cdot28\ \varphi$) and poorly sorted (mean sorting value of 18 samples $= 4\cdot30\ \varphi$, s.d. $0\cdot79$). All samples analysed fell within the very poorly sorted or extremely poorly sorted categories, suggested short-distance transport *en masse*, but with increasing distance from possible source areas, there is a tendency for the head to become finer and less stony as some comminution of debris occurs and some of the coarser blocks are left behind. All samples fell within the range of frost-heaving materials and were characterized by low liquid limits, such that only 25 per cent of water by weight of sediment would suffice to liquefy it and cause flow.

5.1   Talus

Talus or scree is familiar in all regions subjected to frost if suitable rock is available. Under a severe periglacial climate, talus accumulations may reach impressive proportions. A good example is in the Devil's Lake region of the so-called Driftless Area of Wisconsin. On the flanks of the depression containing Devil's Lake, talus slopes rise 100 m in places at angles up to 36°, consisting of massive angular blocks of the purple Baraboo quartzite firmly wedged together. Some blocks are 3 m in length; they have clearly fallen from rock ledges above after frost splitting. The talus is now inactive—lichens grow on the rocks, even trees in places; new additions of rock from the outcrops above have virtually ceased, and no injured trees are to be seen. The talus clearly accumulated under much more severe conditions than those of today, and probably most of it at a time when Devil's Lake was formed by late Wisconsin ice plugging the gap to the south-east. The talus in fact continues at its foot beneath till of Cary age.

In an area of greatly contrasting present-day climate, R. W. Hey (1963) describes

Plate X
Talus slopes at foot of basalt escarpment, Breiðamerkurjökull, Iceland. (C.E.)

the occurrence of Pleistocene screes in North Africa. Angular limestone debris in Cyrenaica forms giant fossil scree-slopes with gradients of up to 30°, now partly vegetated. Chronological links with late Pleistocene cold periods are based on correlations with terrace gravels and included Palaeolithic artefacts.

In regions of present-day periglacial climate, talus accumulation is active. Rapp (1960a, 1962) has studied processes of talus movement in Kärkevagge, northern Sweden. On an active talus cone with slopes up to 38°, there is slow downslope creep amounting to 10 cm a year at a maximum, caused by freeze-thaw of interstitial ice, and by settling of the blocks owing to water washing out fines. Creep is active because of the cold humid climate and the rich supply of mobile schistose debris. Fragments also roll and slide down the surface when they fall from above, and small slides occur whenever equilibrium is disturbed. The total displacement of material on talus slopes in the area may amount to as much as 5000 t/m per year, though this is small in comparison with other processes of mass movement in Kärkevagge, notably boulder falls, avalanches and mud slides. Rapp (1960b) has also studied modern frost-climate screes in Spitsbergen; at Tempelfjorden, there are on average 59 days a year with complete frost cycles of up to 5·5°C amplitude, confined to spring and autumn. Scree formation ceases in winter when conditions are too cold.

5.2   *Stratified slope deposits*

Periglacial slope deposits consisting of alternating beds of fine and coarse debris are sometimes known as 'grèzes litées' (Y. Guillien, 1951) or 'éboulis ordonnées' (J. Tricart, 1953, 1956). The fine-grained layers are often silty, while the coarser layers, as well as being thicker, may contain fragments 10 cm or more in length. In Lorraine, grèzes litées attain thicknesses of 40 m. The beds tend to thicken downslope, so that their dip often increases with depth, up to 27° in west-central Wales (E. Watson, 1965) and up to 24° in the southern French Alps (Tricart, 1953). In the Southern Alps of New Zealand, J. M. Soons (1961) has recorded slope and dip angles of 30–35°. Some splitting or joining of beds occurs; cryoturbation structures are rarely seen in the Welsh examples, but are noted, for instance, by T. Czudek (1964) in grèzes litées in Czechoslovakia. J. Dylik (1960) has surveyed the occurrence of grèzes litées in Poland and finds them thinner and containing more fine material than their French equivalents. This difference he relates to the more continental climate of Poland and to the fact that, unlike France, glacial deposits supplied a great deal of fine material.

There is much uncertainty about their origin. The main problem is to account for their stratification. Basically, the deposits as a whole are built out of frost-broken fragments. It is possible that the layers of coarser fragments accumulate under cold conditions (either seasonally or over longer periods) and are spread out over the surface of the scree by sliding on a frozen or snow-covered surface. A. Cailleux (1963) points out that permafrost is not needed—seasonal frost or just snow would suffice, and he notes Jahn's observations in Spitsbergen of coarse gravelly material accumulating on snow slopes. The fine-grained layers may represent either accumulation under different conditions (for instance, slope wash in the presence of snow-melt) or sorting of the material following its descent to the scree. Dylik (1969) emphasizes the role of slope wash which, he contends, is a vital but often underestimated factor in periglacial slope development. A frozen sub-surface prevents downward percolation; the thawed crust of the deposits in summer or warm intervals liberates meltwater flowing as a sheet and carrying with it the finer particles to be deposited as a thin discrete layer. Under conditions of more prolonged thawing, furrows may be eroded in the deposits, often along lines of ice veins or wedges. Such furrows, later infilled, are often encountered in sections of these stratified deposits (e.g., Guillien, 1964). Soons (1961) suggests that needle-ice may play a part in the sorting while Guillien (1964) points out that a covering of snow on the slope may act as a sort of filter through which only fine particles may be washed by water soaking through the snow from surface melt. Studies have been made of the orientation of stratified screes in the hope that some possible relationships with climatic factors, including the incidence of snow, might be established, but no useful generalizations have yet been established. In Lorraine, grèzes litées occur mostly on east or north-east facing slopes, where most permanent or seasonal snow beds would lie, but in Charente, most grèzes litées face south-east, and of the fourteen sites studied by Watson in west central Wales, seven face south-east, four south-west, and none faces north-east. Further studies are clearly needed on the origin of these deposits.

6 Fracture and deformation of bedrock by periglacial mass movement

Deep ground freezing is competent to widen existing bedrock joints and also to open up bedding planes. On slopes, whole joint blocks may become separated and take part in downhill movement. Many classic examples of the process, active in the Pleistocene, are to be found along the escarpment of Niagaran dolomite in Ontario and Wisconsin. On the flanks of the Platteville mounds (Plate XIA), outliers of the Niagaran formation, huge blocks of dolomite now partially encased in modern soil and vegetation, can be seen in every state of tilting and downslope collapse. A. Straw (1966) has discussed the phenomenon in part of Ontario. The fissures that form during the initial stages of blocks breaking away from the edge of the main outcrop are known as gulls; with the ending of periglacial conditions and the cessation of activity, these become infilled, but often only partially, so that the surface expression of cracks trending parallel to the surface contours remains. A related phenomenon, first described from the North-ampton Ironstone Field of central England by S. E. Hollingworth, J. H. Taylor and J. A. Kellaway (1944) is that of cambering, the convex bending of otherwise horizontal or near-horizontal strata at valley margins. This may be partly a thermokarst effect, associated with the melting of ground ice in an incompetent underlying formation: cambering and sagging of stronger overlying beds will then result. (Note that Hollingworth et al. attributed the cambering of Jurassic rocks to plastic deformation of underlying clays due to differential unloading during valley cutting; and it should not be thought that cambered structures are necessarily all periglacial.)

Plate XIA (*above opposite*)
Mass movement, under former periglacial conditions, of large blocks of Niagaran dolomite, near Platte-ville, Wisconsin. (C.E.)

Plate XIB (*below opposite*)
Gelifluction lobes in Baffin Island, Arctic Canada. (C.A.M.K.)

7 Morphological effects

By stripping off rock waste from higher ground and redepositing it lower down, especially in valley floors, gelifluction and other slope processes can slowly reduce the amplitude of relief; at the same time, smoothing of the landscape is effected by burial of minor bedrock irregularities. The concept of landscape levelling under a periglacial climate was termed 'equiplanation' by D. D. Cairnes (1912) though he included processes other than gelifluction. Bryan (1946) offered the term 'cryoplanation' (Chapter 6) to indicate general reduction of the land surface by periglacial weathering and mass movements, and included the work of rivers (often flowing only seasonally, as in parts of east Greenland and Spitsbergen described by H. Poser, 1936) in transporting material delivered to the valley floors by downslope movement.

Wherever there are changes in the rates of movement downslope, certain minor land-forms such as steps or banks will result. These have already been briefly described in Chapter 3, since they contribute to the general category of patterned ground. The following is Lundqvist's classification (1949):

	Ground rich in boulders	Few boulders
Rich cover of vegetation	Stone-banked lobes and terraces	Turf-banked lobes and terraces
Little or no vegetation	Stone streams	Earth wrinkles

The difference between lobes and terraces is purely morphological: terrace edges are more linear and may follow or run slightly obliquely to the contour, while lobes have a small extent along the contour in comparison with their downslope dimension. Stone streams will be dealt with in Chapter 6 during consideration of the related forms of blockfields and boulder fields.

7.1   Stone-banked lobes and terraces

These comprise gelifluction and other deposits confined by crescent-shaped stony embankments (the *Steinguirlanden* of Högbom, 1914, p. 335), and stone-banked terraces. They occur on moderate slopes, usually 10–25°. Their size is variable: Jahn (1958) describes examples from the Tatra Mountains ('solifluial tongues') which have frontal banks 1–4 m high; R. P. Sharp (1942) in the St. Elias mountains of Yukon notes stone garlands consisting of stony embankments up to 1 m high and extending up to 8 m downslope (Fig. 4.6); Galloway (1961) quotes stone-banked lobes and terraces from the Scottish Highlands with risers as much as 5 m high and treads extending up to 30 m from front to back (Fig. 4.7). The treads of the terraces and the central parts of the lobes, underlain by relatively fine-grained material, usually slope gently outwards at angles as low as 2° or 3° with gentle undulations and sometimes frost hummocks on their surfaces. The formation of the stony embankments is incompletely understood. Stones may be heaved to the surface by frost and then move downhill faster than the slow flowage of the fine materials, according to Sharp, and in support of this, Jahn

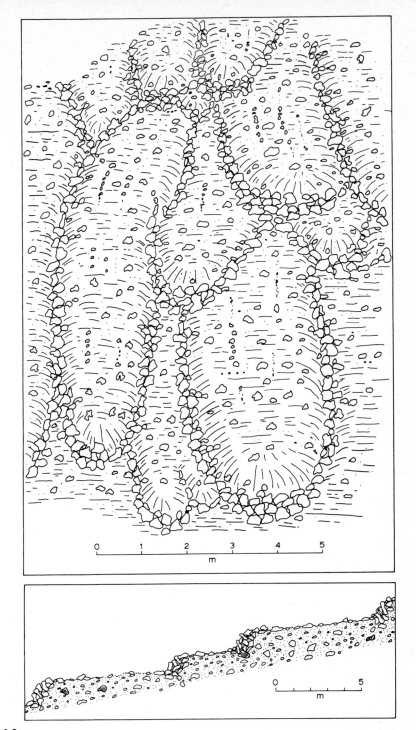

Fig. 4.6
Stone-banked terraces and garlands; plan and section (R. P. Sharp, *J. Geomorph.*, 1942, by permission of the author)

(1958) draws attention to the fact that large stones on the tongue surfaces often have small turf or earth wrinkles in front of them (see discussion of 'ploughing blocks' on p. 118). Taber (1943) also thought that large stones move faster on a slope, sometimes by rolling or overturning. The downward movement of the stones is arrested by slight

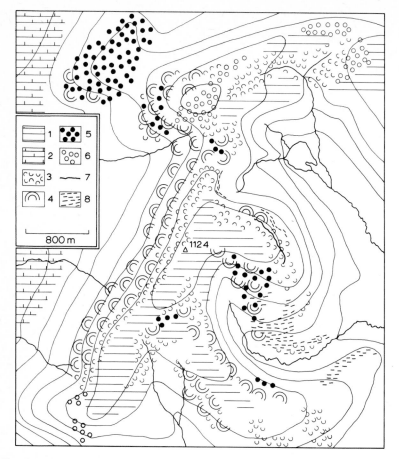

Fig. 4.7
Gelifluction and related phenomena on Ben Wyvis, Scotland (R. W. Galloway, *Scott. geogr. Mag.*, 1961, Royal Scottish Geographical Society).
1 partly active gelifluction sheets on plateau; 2 inactive lower level gelifluction sheets; 3 active turf-banked lobes; 4 inactive stone-banked lobes; 5 blockfields and stone streams (mostly inactive); 6 ring and stripe patterns; 7 continuous turf-banked terrace (inactive); 8 discontinuous turf-banked terraces (active).

changes in slope, by clumps of vegetation, by bedrock exposures, or by simple accumulation in which several stones catch up with one another and jam to form the nucleus of a growing pile. When the stone bank becomes large enough, little further movement apart from creep will occur. It then grows in height as more material is added. J. B. Benedict (1966) has been able to estimate rates of motion at different periods in the last 2500 years for stone-banked terraces in the Colorado Front Range. The terraces

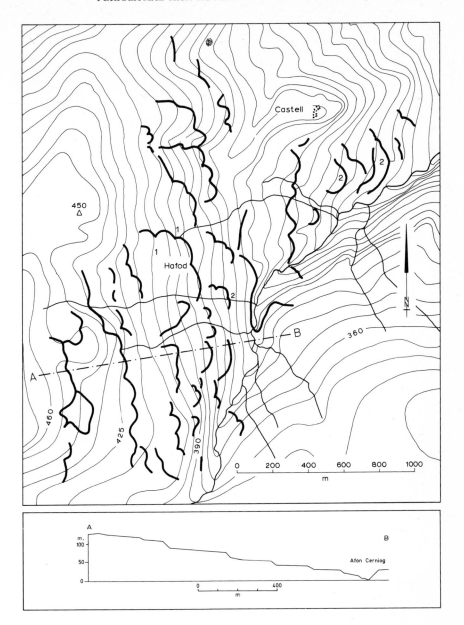

Fig. 4.8
Turf-banked terraces in the Afon Cerniog valley. The profile is located along the line A–B (A. Pissart, *Biul. Peryglac.*, 1963)

were overriding humus layers, and radiocarbon dates from different parts of one such layer were obtained. These gave rates of downslope movement of the terrace edge ranging from less than 1·5 mm/year to a maximum of 23 mm/year. Jahn (1967) observes that gelifluction lobes and tongues may become stabilized, or their movement retarded, if runoff begins to wash out the fines.

7.2 *Turf-banked lobes and terraces*

In shape and size these differ little from the stone-banked variety, though there seems to be a tendency for them to occur on slightly more gentle slopes (5–20°). A. Pissart (1963) has identified lobate turf-covered terraces in central Wales which he believes have formed by gelifluction and frost creep below transverse snow-banks persisting on north-east slopes. Preferred orientation of stones in the terraces gives clear evidence of downslope movement. Terrace fronts are from 2–15 m high, their treads extend for 50–500 m, and they occur on slopes averaging 2·5–8° (Fig. 4.8). Examples described by Galloway (1961) in the Scottish Highlands are much smaller (risers up

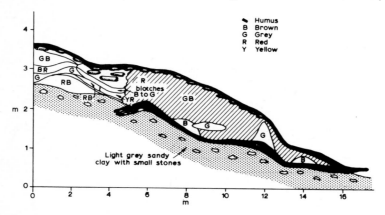

Fig. 4.9
Downslope section of a terrace on the south slope of Blåho, Trollheimen, Norway (P. J. Williams, *Geogrl J.*, 1957, Royal Geographical Society)

to 1·5 m high, treads 2–6 m wide). There is a clear evidence in some areas that turf-banked terraces, like the stone-banked variety, are actively in motion. The turf cover may be ruptured in places, releasing a flow of fine debris which spreads fan-like on to the next lower terrace. Williams (1957) notes how the bulging fronts of some turf-banked terraces (Fig. 4.9) on Dovrefjell, Norway, are overriding vegetation and old humus layers. One such layer was traced beneath an overriding terrace for 13 m, and using rough estimates of the maximum age of the layer, the rate of overriding may have averaged 2 mm a year. In areas where both stone-banked and turf-banked terraces occur (for example, on Niwot Ridge, Colorado: J. B. Benedict, 1966, 1970), the former seem to move somewhat faster downslope than the latter, which are overridden by the stones. The materials in the terraces are highly contorted in section, and Troll (1944) shows examples of recumbent folds in soil horizons of terraces in the Hohe Tauern.

Benedict (1970) has measured average rates of movement of stakes in active terraces in the Colorado Front Range (Fig. 4.10). For terrace 19, the maximum movement for the period 1964–7 was at a centrally placed stake, amounting to 10·9 mm/year. Fabric studies in the moving material of the terraces (Fig. 4.11) showed that elongate sand particles are oriented parallel to the long axes of cobbles and boulders and to the direction of slope. Dips vary from horizontal to up-slope, and the fabric becomes increasingly imbricated as the front of the lobe is approached. Radiocarbon dates from buried organic horizons were obtained, enabling a history of past downslope movement to be reconstructed.

The terraces may develop in areas of either complete or patchy turf cover. Patchiness

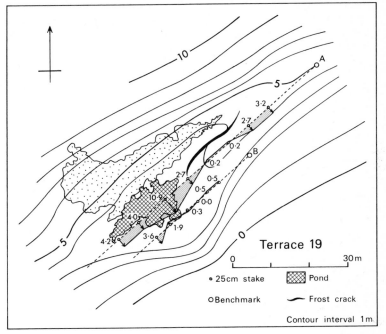

Fig. 4.10
Map of a turf-banked terrace in the Colorado Front Range, showing average rates of movement (mm/year) for the period 1964–7. The front of the terrace, picked out by the contours, bulges downslope below an ephemeral pond fed by a late-lying snow-bank (stippled). The position of the snow-bank is that of 8 July 1965. (J. B. Benedict, *Arctic and Alpine Research*, 1970)

may be natural, or it may reflect the activities of man or animals, as in the Tatra Mountains (Jahn, 1958). Destruction of the turf enables frost to penetrate more deeply, and facilitates washing of fines by any surface runoff. Terrace surfaces may support small pools of water at times, encouraging comminution of the sub-surface materials by frost. Patterned ground is commonly observed on them where the vegetative cover is limited.

On the fronts of lobes or terraces, the turf serves to restrain movement as in the case of stony embankments. The variations in its restraining power on different angles of slope are well illustrated by S. E. Hollingworth (1934) (Fig. 4.12). In north-west Scotland, Mottershead and I. D. White (1969) stress the importance of the inter-relationships between the development and form of the terraces, and the plant cover.

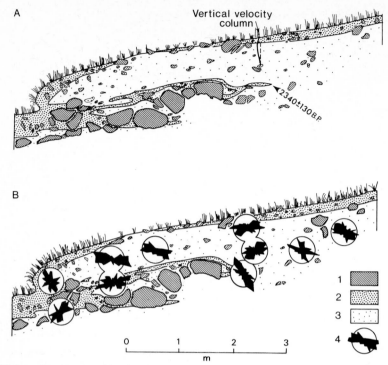

Fig. 4.11
Soil profile through lobe 499, Colorado Front Range (J. B. Benedict, *Arctic and Alpine Research*, 1970).
1 Stones, drawn to scale.
2 Very dark brown/greyish-brown gravelly sandy loam.
3 Dark greyish-brown/yellow-brown sandy loam, cohesive, with weak platy structure.
4 Diagrams showing the long axis orientations of elongate sand grains. Each diagram summarizes 200 measurements; circles represent 10 per cent frequencies

Three main types of terrace are distinguished: the large stone-banked variety with relatively flat treads and low-angle risers between the terraces; moderate-sized turf-banked terraces whose risers may attain angles of 30–60° but whose treads, sparsely vegetated, are still gently inclined; and turf lobes, where both facets are steeper and well vegetated. Unlike those studied by Benedict at high levels in Colorado, these terraces and lobes in north-west Britain show little sign of activity today. It is interesting that particle-size analysis of the material in them shows that it is mostly too coarse to heave except in the lowest horizons. Possibly fines have been leached out since the terraces were formed, resulting in their virtual stabilization under present climatic conditions.

7.2 *Ploughing blocks*

The term is applied to 'those blocky components of periglacial slope movement which travel faster than their surroundings and which force up the ground before them and leave a depression to their rear' (L. Tufnell, 1972, p. 240). An early reference to this

phenomenon is by Högbom (1914) to 'gleitende Blöcke'. Rates of movement are probably very variable: sudden slippage may be involved as well as longer-period slow
creep. Tufnell suggests several possible causes of movement. Temperature change causing expansion and contraction of the block on a slope will tend to result in net downhill

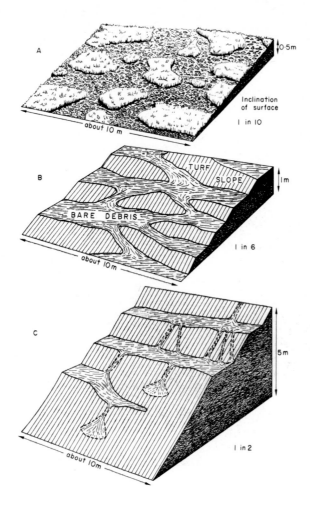

Fig. 4.12
Diagrams illustrating the patterns formed by moving debris on partially turfed slopes of various inclina-
tions (S. E. Hollingworth, *Proc. Geol. Ass.*, 1934)

creep, as will frost-heaving. Snow or meltwater accumulating upslope from the block
will aid lubrication of the base of the block and soak the ground around it—snow often
disappears first from below the downslope edge due to meltwater dripping off the lower
edge of the block. Possibly the commonest cause of block movement is when the block
is located in an active layer above permafrost.

Plate XIIA
Lower part of the Arapaho rock glacier rising above the snow in the foreground, Colorado Front Range. (C.E.)

Plate XIIB
View of Arapaho rock glacier from above. (C.A.M.K.)

8 Rock glaciers

The term rock glacier was first used by S. R. Capps (1910) in Alaska. In the Kennicott region he investigated thirty or so rock glaciers built of angular talus, extending up to the cirque headwalls and showing no ice at the surface; but excavations revealed that ice filled the interstices between the blocks at depths within a metre of the surface. In some cases they grade into true glaciers at their upper ends. D. Barsch (1969) distinguishes several types of rock glacier including those that originate at the base of talus slopes, those that extend down-valley from large glacial end-moraines, and those that are really true glaciers thickly covered with rock rubble. The distinction between rock glaciers and ice-cored moraines has also proved difficult and controversial, Barsch (1971) claiming that the terms are synonymous and G. Østrem (1971) contesting this view (for further discussion, see Embleton and King, 1975, p. 439).

Microrelief on the surfaces of rock glaciers, especially lobate wrinkles and ridges at their lower ends, often gives a vivid impression of motion, yet it has proved to be exceedingly slow motion, and some rock glaciers are now stagnant, their fronts covered with turf and lichen. Rock glaciers are known from several parts of North and South America and Europe: E. Howe (1909) described their forms in the San Juan mountains

(Colorado) but thought, wrongly, that they were landslide debris; A. Chaix (1923) measured their movements in the Engadine (Switzerland); R. L. Ives (1940) in the Front Range of Colorado (Plates XIIA and B), J. E. Kesseli (1941) in the Sierra Nevada, and Mlle A. Faure-Muret (1949) in the French Alps, added to the available observations. Some of the largest recorded by Kesseli attain lengths of 3 km with terminal embankments 60 m high. Component blocks up to 8 m in size are known. There is general agreement that rock glaciers have an upper crust of angular blocks without interstitial debris (though there may be ice) resting on a much thicker lower layer of angular blocks, sand, silt, and possibly mud (again, with or without ice). This differentiation into two layers is caused by downward sifting of fines, and to the greater amount of grinding to which the blocks in the basal layers are subjected.

A recent thorough study of Alaskan rock glaciers by C. Wahrhaftig and A. Cox (1959) makes it unnecessary to consider all the numerous theories of origin of rock glaciers that have been expressed. Their work shows plainly that the motion of rock glaciers is the flow of frozen rubble akin to glacier flow. The first accurate measurements of rock glacier motion were made by Chaix (1923 and 1943), who over a 24-year period showed that, at the terminus of the Val Sassa rock glacier (Upper Engadine), the surface had moved at an average rate of 1–1·5 m/year compared with 0·3–1 m/year at the base of its terminal face. Clearly, the upper layer was moving faster than the lower parts.

Wahrhaftig and Cox measured the surface motion of one Alaskan rock glacier over 8 years (maximum rate 0·76 m/year). In this example, the rock glacier was about 30 m thick and all but the upper 2 or 3 m was perennially frozen. Hence ice must participate in its motion. Since the talus aprons at its front built up to about two-fifths of the height of the front, it was possible to obtain an approximate idea of the vertical distribution of velocity (Fig. 4.13). Such a flow would exist in a material having a strain rate proportional to stress, again suggesting that flow of interstitial ice was taking place. Supporting this were estimates of the shear stresses involved. Using the approximate relationship $\tau = \rho g h \sin \alpha$, where τ is the maximum shear stress, ρ = density (taken as 1·8 g/cm³), h = flow thickness and α = surface slope, they found a range of values for 30 rock glaciers from less than 1 to about 2 bar, similar to values obtained for true glaciers. Approximate calculations were also made of rock glacier viscosity (η):

$$\eta = \rho g h^2 \sin \alpha / 2v$$

where v = mean surface velocity and other quantities are as previously defined. It was assumed that velocity at the base of the flows was zero. For three rock glaciers, viscosity varied between 1·6 and 9·0 poises $\times 10^{14}$. These figures are slightly greater than those normally associated with glacier ice (10^{12}–10^{14} poises) which is to be expected since rock glaciers contain high proportions of rigid detritus. Rates of movement measured by S. E. White (1971) for three rock glaciers in Colorado were an order of magnitude less than those measured by Wahrhaftig and Cox. Average rates computed for 10–16 markers on each rock glacier amounted to only 5·0, 6·6 and 9·7 cm/year respectively. Estimated values of viscosity were much higher than those obtained by Wahrhaftig and Cox: 90, 54 and 92 poises $\times 10^{14}$ respectively, but theoretical maximum shear stress values near the rock glacier fronts were similar, ranging from 0·97 to 1·35 bar.

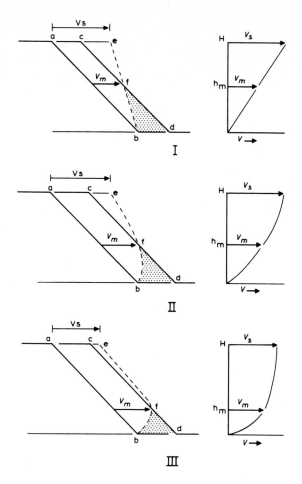

Fig. 4.13
Three possible vertical velocity profiles (on right of diagram) for a rock glacier, as determined from the position of the top of the talus apron at the rock glacier front (C. Wahrhaftig and A. Cox, *Bull. geol. Soc. Am.*, 1959).

$\left.\begin{matrix} ab \\ cd \end{matrix}\right\}$ successive positions of the rock glacier front

V_s surface velocity

V_m velocity of rock debris (at point f, this equals the velocity of the front. Thus the areas cef and bdf are equal in all three cases)

   f   position of the top of the talus apron

Diagram II is the most common situation (talus apron rising 30–45 per cent of the total height of the front), and diagram III represents the next most common. Conditions of diagram I (h_m at half the total height of the front) are less common. The flow diagram in situation II would characterize a material having a strain rate proportional to stress (*cf.* glacier ice).

The existence of ice in the lower layers of active rock glaciers is therefore almost certain, though obviously difficult to prove by direct observation. Unique in this respect were W. H. Brown's observations (1925) of the internal structure of a rock glacier in Colorado from a mining tunnel dug in from its lower end. This passed first through loose rock, then for some distance through the rock glacier with interstitial ice everywhere, finally passing through a small quantity of glacial ice before entering solid rock. In the absence of tunnels, drill holes provide the only other means of internal investigation, yet the combination of ice, rock fragments and boulders makes this virtually impracticable. White (1971) used dynamite to blast a hole into Arapaho rock glacier in the Colorado Front Range, confirming that it has a frozen core. The hole revealed a matrix of muddy sand and 1–2 cm ice crystals.

N. Potter (1972) has recently contested the view of Wahrhaftig and Cox that rock glaciers normally consist dominantly of rock debris with interstitial ice. The Galena Creek rock glacier in Wyoming turns out to have a maximum debris content of only 10–12 per cent by volume—it is approaching the condition of a normal 'dirty' glacier with a thick skin of rubble. Potter proposes that there are two main classes of rock glacier, in a continuum from clean glaciers at the one extreme to rock glaciers consisting mainly of debris and little ice at the other. These two intermediate classes are:

1 ice-cored rock glaciers (such as the Galena Creek example)
2 ice-cemented rock glaciers (such as those described by Wahrhaftig and Cox).

It is admitted, however, that evidence is often insufficient to permit classification in this way. Potter suggests that, in class 1, the debris mantle acts as a very good insulator so that, characteristically, this sort of rock glacier has a large ablation area compared with its accumulation area (in the case of the Galena Creek rock glacier, the ratio is 7 : 1).

Microrelief on rock glacier surfaces is analysed comprehensively by Wahrhaftig and Cox. Longitudinal furrows appear to result from wasting of ice-rich bands which develop in the gaps between talus cones feeding the head of a rock glacier. Meltwater in the furrows deepens them further. Transverse ridges, conveying such a strong impression of glacier-like flow are thought to represent wrinkling of the surface crust or even internal shearing as the motion of the rock glacier is slowed; they are probably not annual features such as glacier ogives. Conical pits represent meltwater sinks.

Rates of erosion can be roughly computed for rock glaciers. They are fed by frost-split rocks from surrounding slopes, so that source areas can be defined. The amount of vertical rock removal is assessed by computing the volume of rock debris in a rock glacier and dividing it by the source area. In examples studied by Wahrhaftig and Cox, the rock glaciers have probably been active for little more than 3000 years, the period since true glacial ice last occupied the cirques, and a rate of denudation amounting to 3 m (or even 6–9 m) per 1000 years seems likely. White (1971) uses data on average rates of movement, thickness and width to obtain current rates of debris removal for three rock glaciers in Colorado of 215, 269 and 771 m³/year respectively.

9 Conclusion

Two main processes of periglacial downhill movement, formerly termed solifluction, are identified:

1 gelifluction, the flow of water-soaked debris resulting from seasonal thawing of the active layer;
2 frost creep, caused by alternate freezing and thawing of slope deposits.

In the first case, an excess of water reduces the shear strength of the material, and relatively rapid movements may result if vegetation is unable to prevent them. In the second case, frost-heave produces planes of weakness within the material and reduces cohesion between the particles. On thawing, the material resettles downhill. This process is only significant in frost-susceptible materials. Rates of movement by creep have been over-estimated in the past and are more likely to be measured in centimetres rather than metres per year; on the other hand, rates of movement by flow, sporadic and localized, may amount to several metres in a few days.

Gelifluction and related deposits show evidence of flow orientation and contorted structures caused by movement. Stone-banked and turf-banked lobes and terraces are common morphological features resulting from differential rates of movement on slopes of 5–25°.

Other periglacial slope deposits include talus and stratified screes. The origin of the stratification in the latter is not clearly understood. A final feature discussed under the general heading of periglacial mass movements is the rock glacier. Most rock glaciers are now known to consist of frozen rubble whose flow is akin to that of glaciers. Rates of surface motion are usually less than 1–2 m/year, slower than in the case of true glaciers because of the extra rigidity imposed by the high proportion of rock debris. Their surfaces possess a unique microrelief related both to movement and the melting of internal ice.

10 References

ANDERSSON, J. G. (1906), 'Solifluction, a component of sub-aerial denudation', *J. Geol.* **14**, 91–112

BAIRD, P. D. and LEWIS, W. V. (1957), 'The Cairngorm floods, 1956: summer solifluction and distributary formation', *Scott. geogr. Mag.* **73**, 91–100

BARSCH, D. (1969), 'Studien und Messungen an Blockgletschern in Macun, Unterengadin', *Z. Geomorph., Suppl.* **8**, 11–30

(1971), 'Rock glaciers and ice-cored moraines', *Geogr. Annlr* **53**A, 203–6

BAULIG, H. (1956a), *Vocabulaire franco-allemand de géomorphologie* (Paris)

(1956b), 'Pénéplaines et pédiplaines', *Soc. belge Études géogr.* **25**, 25–58

BENEDICT, J. B. (1966), 'Radiocarbon dates from a stone-banked terrace in the Colorado Rocky Mountains, U.S.A.', *Geogr. Annlr* **48**, 24–31

(1970), 'Downslope soil movement in a Colorado Alpine region: rates, processes and climatic significance', *Arct. alp. Res.* **2**, 165–226

BROSCOE, A. J. and THOMSON, S. (1972), 'Observations on an alpine mudflow, Steele

Creek', *Icefield Ranges Res. Proj. scient. Results* (ed. V. C. BUSHNELL, and R. H. RAGLE, *Am. Geogr. Soc.* and *Arctic Inst. N. America*) **3**, 53–60

BROWN, W. H. (1925), 'A probable fossil glacier', *J. Geol.* **33**, 464–6

BRYAN, K. (1946), 'Cryopedology—the study of frozen ground and intensive frost action with suggestions on nomenclature', *Am. J. Sci.* **244**, 622–42

BÜDEL, J. (1937), 'Eiszeitliche und rezente Verwitterung und Abtragung im ehemals nicht vereisten Teil Mitteleuropas', *Petermanns Mitt., Ergänz.* **229**, 71 pp.

   (1944), 'Die morphologischen Wirkungen des Eiszeitklimas im gletscherfreien Gebiet', *Geol. Rdsch.* **34**, 482–519

CAILLEUX, A. (1963), 'Processus supranivaux et grèzes litées', *Biul. Peryglac.* **12**, 145

CAILLEUX, A. and TRICART, J. (1950), 'Un type de solifluction: les coulées boueuses', *Revue Géomorph. dyn.* **1**, 4–46

CAIRNES, D. D. (1912), 'Differential erosion and equiplanation in portions of Yukon and Alaska', *Bull. geol. Soc. Am.* **23**, 333–48

CAPPS, S. R. (1910), 'Rock glaciers in Alaska', *J. Geol.* **18**, 359–75

   (1919), 'The Kantishna region, Alaska', *U.S. geol. Surv. Bull.* **687**, 1–116

CHAIX, A. (1923), 'Les coulées de blocs du Parc National Suisse d'Engadine', *Le Globe, Genève* **62**, 1–38

   (1943), 'Les coulées de blocs du Parc National Suisse—nouvelles mesures et comparison avec les "rock streams" de la Sierra Nevada de Californie', *Le Globe, Genève* **82**, 121–8

CONOLLY, A. P., GODWIN, H. and MEGAW, E. M. (1950), 'Studies in the Post-glacial history of British vegetation: XI. Late-glacial deposits in Cornwall', *Phil. Trans. R. Soc.* **234**B, 397–469

CZUDEK, T. (1964), 'Periglacial slope development in the area of the Bohemian massif in northern Moravia', *Biul. Peryglac.* **14**, 169–93

DAVISON, C. (1889), 'On the creeping of the soil-cap through the action of frost', *Geol. Mag.* **6**, 255–61

DE LA BECHE, H. T. (1839), 'Report on the geology of Cornwall, Devon and West Somerset', *Mem. geol. Surv. Gt Br.*

DINES, H. G., HOLLINGWORTH, S. E., EDWARDS, W., BUCHAN, S. and WELCH, F. B. A. (1940), 'The mapping of head deposits', *Geol. Mag.* **77**, 198–226

DYLIK, J. (1951), 'Some periglacial structures in Pleistocene deposits of Middle Poland', *Bull. Soc. Lettr. Łodz* **3**, 2

   (1960), 'Rhythmically stratified slope waste deposits', *Biul. Peryglac.* **8**, 31–41

   (1967), 'Solifluxion, congelifluxion and related slope processes', *Geogr. Annlr* **49**, 167–77

   (1969), 'Slope development under periglacial conditions in the Łodz region', *Biul. Peryglac.* **18**, 381–410

EMBLETON, C. and KING, C. A. M. (1975), *Glacial geomorphology*

FAURE-MURET, A. (1949), 'Les "rock streams" ou "pseudo-moraines" du Massif de l'Argentera-Mercantour', *Bull. Soc. géol. Fr.* **5**, 19, 118–20

GALLOWAY, R. W. (1961), 'Solifluction in Scotland', *Scott. geogr. Mag.* **77**, 75–87

GUILLIEN, Y. (1951), 'Les grèzes litées de Charente', *Revue Géogr. Pyrénées S.-Ouest* **22**, 154–62

(1964), 'Grèzes litées et bancs de neige', *Geol. Mijnbouw* **43**, 103–12

HAMELIN, L.-E. and COOK, F. A. (1967), *Illustrated glossary of periglacial phenomena* (Quebec)

HARRIS, C. (1972), 'Processes of soil movement in turf-banked solifluction lobes, Okstindan, northern Norway', *Inst. Br. Geogr. Spec. Publ.* **5**, 155–74

HEY, R. W. (1963), 'Pleistocene screes in Cyrenaica (Libya)', *Eiszeitalter Gegenw.* **14**, 77–84

HIGASHI, A. and CORTE, A. E. (1971), 'Solifluction: a model experiment', *Science* **171**, 480–2

HÖGBOM, B. (1914), 'Über die geologische Bedeutung des Frostes', *Bull. geol. Instn Univ. Upsala* **12**, 257–389

HOLLINGWORTH, S. E. (1934), 'Some solifluction phenomena in the northern part of the Lake District', *Proc. Geol. Ass.* **45**, 167–88

HOLLINGWORTH, S. E., TAYLOR, J. H. and KELLAWAY, J. A. (1944), 'Large-scale superficial structures in the Northampton Ironstone Field', *Q. J. geol. Soc. Lond.* **100**, 1–44

HOWE, E. (1909), 'Landslides in the San Juan Mountains, Colorado', *U.S. geol. Surv. Prof. Pap.* **67**

IVES, R. L. (1940), 'Rock glaciers in the Colorado Front Range', *Bull. geol. Soc. Am.* **51**, 1271–94

JAHN, A. (1956), 'Some periglacial problems in Poland', *Biul. Peryglac.* **4**, 169–94

(1958), 'Periglacial microrelief in the Tatras and on the Babia Góra', *Biul. Peryglac.* **6**, 227–49

(1960), 'Some remarks on the evolution of slopes on Spitsbergen', *Z. Geomorph., Suppl. Band* **1**, 49–58

(1967), 'Some features of mass movement on Spitsbergen slopes', *Geogr. Annlr* **49**, 213–25

KESSELI, J. E. (1941), 'Rock streams in the Sierra Nevada, California', *Geogrl Rev.* **31**, 203–27

LUNDQVIST, G. (1949), 'The orientation of the block material in certain species of flow earth', *Geogr. Annlr* **31**, 335–47

MACKAY, J. R. (1972), 'The world of underground ice', *Ann. Ass. Am. Geogr.* **62**, 1–22

MOTTERSHEAD, D. N. (1971), 'Coastal head deposits between Start Point and Hope Cove, Devon', *Field Stud.* **3**, 433–53

MOTTERSHEAD, D. N. and WHITE, I. D. (1969), 'Some solifluction terraces in Sutherland', *Trans. bot. Soc. Edinb.* **40**, 604–20

ØSTREM, G. (1971), reply to BARSCH, D. *Geogr. Annlr* **53**A, 207–13

PATERSON, T. T. (1951), 'Physiographic studies in north-west Greenland', *Meddr Grønland* **151**, 4, see pp. 11 and 27

PISSART, A. (1963), 'Des replats de cryoturbation au Pays de Galles', *Biul. Peryglac.* **12**, 119–35

POSER, H. (1936), 'Talstudien aus Westspitsbergen und Ostgrönland', *Z. Gletscherk.* **24**, 43–98

POTTER, N. (1972), 'Ice-cored rock glacier, Galena Creek, northern Absaroka mountains, Wyoming', *Bull. geol. Soc. Am.* **83**, 3025–58

RAPP, A. (1960a), 'Recent development of mountain slopes in Kärkevagge and surroundings, northern Scandinavia', *Geogr. Annlr* **42**, 65–200

(1960b), 'Talus slopes and mountain walls at Tempelfjorden, Spitsbergen', *Norsk Polarinst. Skr.* **119**, 96 pp.

(1962), 'Kärkevagge: some recordings of mass movements in the northern Scandinavian mountains', *Biul. Peryglac.* **11**, 287–309

REID, C. (1887), 'On the origin of the dry chalk valleys and of the coombe rock', *Q. J. geol. Soc. Lond.* **43**, 364–73

RUDBERG, S. (1962), 'A report on some field observations concerning periglacial geomorphology and mass movement on slopes in Sweden', *Biul. Peryglac.* **11**, 311–23

SCHUBERT, C. (1972), 'Striated ground in the Venezuelan Andes', *J. Glaciol.* **12**, 461–8

SHARP, R. P. (1942), 'Soil structures in the St. Elias Range, Yukon Territory', *J. Geomorph.* **5**, 274–301

SIGAFOOS, R. S. and HOPKINS, D. M. (1952), 'Soil instability on slopes in regions of perennially frozen ground' in 'Frost action in soils: a symposium', *Highway Res. Board* (Washington)

SMITH, H. T. U. (1949), 'Periglacial features in the driftless area of southern Wisconsin', *J. Geol.* **57**, 196–215

SMITH, J. (1960), 'Cryoturbation data from South Georgia', *Biul. Peryglac.* **8**, 73–9

SOONS, J. M. (1962), 'A survey of periglacial features in New Zealand', in *Land and Livelihood: Geographical Essays in honour of George Jobberns* (ed. M. McCASKILL, *N.Z. geogr. Soc.*, Christchurch), 74–87

STRAW, A. (1966), 'Periglacial mass movement on the Niagara escarpment near Meaford, Grey County, Ontario', *Geogr. Bull.* **8**, 369–76

TABER, S. (1943), 'Perennially frozen ground in Alaska: its origin and history', *Bull. geol. Soc. Am.* **54**, 1433–548

THOMPSON, W. F. (1962), 'Preliminary notes on the nature and distribution of rock glaciers relative to true glaciers and other effects of the climate on the ground in North America', *Un. géod. géophys. int., Symposium at Obergurgl, 1962 (Publication No. 58, Int. Ass. scient. Hydrol.)*, 212–19

TRICART, J. (1953), 'Les actions périglaciaires du Quaternaire récent dans les Alpes du Sud', *Rep. 4th Conf. int. Ass. quatern. Res. (Rome, 1953)*, 189–97

(1954), 'Premiers résultats d'expériences de solifluxion périglaciaire', *C. r. hebd. Séanc. Acad. Sci., Paris* **238**, 259–61

(1956), *Cartes des phénomènes périglaciaires Quaternaires en France. Mémoires* (Paris)

TROLL, C. (1944), 'Strukturböden, Solifluktion, und Frostklimate der Erde', *Geol. Rdsch.* **34**, 545–694 (English translation, *Snow Ice Permafrost Res. Establ.*)

(1947), 'Die Formen der Solifluktion und die periglaziale Bödenabtragung', *Erdkunde* **1**, 162–75

TUFNELL, L. (1972), 'Ploughing blocks with special reference to north-west England', *Biul. Peryglac.* **21**, 237–70

WAHRHAFTIG, C. and COX, A. (1959), 'Rock glaciers in the Alaska Range', *Bull. geol. Soc. Am.* **70**, 383–436

WASHBURN, A. L. (1947), 'Reconnaissance geology of portions of Victoria Island and adjacent regions, Arctic Canada', *Am. geol. Soc. Mem.* **22**

   (1967), 'Instrumental observations of mass wasting in the Mesters Vig district, north-east Greenland', *Meddr Grønland* **166** (4), 296 pp.

   (1973), *Periglacial processes and environments*

WATSON, E. (1965), 'Grèzes litées ou éboulis ordonnés tardiglaciaires dans la région d'Aberystwyth, au centre du Pays de Galles', *Bull. Ass. Géogr. fr.* **338–9**, 16–25

WHITE, S. E. (1971), 'Rock glacier studies in the Colorado Front Range, 1961–68', *Arct. alp. Res.* **3**, 43–64

WILLIAMS, P. J. (1957), 'Some investigations into solifluction features in Norway', *Geogrl J.* **123**, 42–58

   (1959), 'An investigation into processes occurring in solifluction', *Am. J. Sci.* **257**, 481–90

   (1961), 'Climatic factors controlling the distribution of certain frozen ground phenomena', *Geogr. Annlr* **43**, 339–47

5

The action of snow

The daily melting of the snow keeps the rocks around thoroughly soaked with water at zero temperature, so that the action of frost is more than usually effective in splitting them to pieces. (W. B. WRIGHT, 1914)

In considering the geomorphological effects of snow in non-glacierized regions, a distinction may be drawn between snow remaining largely motionless on flat or moderately sloping surfaces, and snow moving rapidly as avalanches on steep gradients. Both aspects will be considered in this chapter.

1 Nivation

The term 'nivation' was introduced by F. E. Matthes (1900) in his study of the Bighorn Mountains, Wyoming, to describe the erosive effects associated with an immobile and patchy snow cover. Since Matthes drew attention to it, nivation has been studied in both alpine and arctic environments, and its effects widely recognized in areas formerly marginal to the Pleistocene ice sheets.

The dominant process involved in nivation is freeze-thaw. The first detailed studies of freeze-thaw action associated with snow patches were undertaken by W. V. Lewis (1936, 1939) in an area near the northern edge of Vatnajökull, Iceland and by L. H. McCabe (1939) in Vest-Spitsbergen. Both were convinced of the importance of freeze-thaw action around and sometimes beneath snow patches but their work showed that the process is by no means a simple one.

In the first place, it is necessary to distinguish between thin snow patches of limited extent, and thicker snow drifts covering larger areas. No precise distinction can be laid down but the important difference is that the thicker snow drift will insulate the ground surface from atmospheric freeze-thaw cycles; the minimum thickness will therefore depend on their duration and amplitude. An additional consideration is whether or not permafrost underlies the snow patch. There are thus four main situations to be considered:

1 Thick snow over permafrost. The ground surface will be subject to frost attack if meltwater generated on the snow surface seeps beneath the snow and, after melting

a thin top layer of the permafrost, refreezes. It should be noted, however, that permafrost will not usually form or survive beneath thick, extensive, and persistent snow cover (Chapter 2).

2 Thick snow over unfrozen ground. The effect will be protective.

3 Thin snow over permafrost. Atmospheric freeze-thaw cycles may alternately cause melting of the snow and upper permafrost, and refreezing.

4 Thin snow over unfrozen ground. The effect will be as in case 3 except that refreezing, unaided by the presence of permafrost below, may be slower.

The snow itself plays a dual role; on the one hand, it is a most important source of meltwater, and on the other it acts as a very efficient insulating medium. Permafrost, if present, acts by providing a reservoir of cold, and by preventing loss by percolation of meltwater from the snow or ground ice.

Since even thin snow will exert some protective influence on the ground beneath in respect of atmospheric freeze-thaw cycles, the zone of maximum destruction by frost action will be around the edges of the snow patch, where meltwater is available but the ground not actually snow covered. Moreover, as the position of the snow margin changes through the year, so this zone of destructive action will migrate with it.

An example of the relationship between air temperature fluctuations and the conditions around and beneath a thin snow patch in Vest-Spitsbergen is given by McCabe (1939) for the night of 31 August 1938. Between 8 p.m. and 3 a.m., the air temperature fell from $+2°C$ to $-3°C$. By this time, the ground around the snow patch had frozen hard to a depth of 2 cm, though below this depth it remained wet and unfrozen down to the permafrost level. Beneath the snow patch, a thin sheet of ice formed, owing to freezing of the previous day's meltwater. The air temperature rose above freezing-point by 10 a.m. but melting of the snow surface did not commence until 2 p.m. when the air temperature was $4°C$.

It should be borne in mind that the average number of days per year when temperatures cross freezing-point is relatively few in Spitsbergen (59 at Green Harbour) and confined to spring and autumn. At high altitudes in tropical regions, however, the number of frost cycles per year is very much greater, exceeding 300 in a few places in the Andes (C. Troll, 1944). Nivation in such areas will be more potent.

Although it has been generally agreed that thick snow patches overlying unfrozen ground will exert a relatively protective effect, the possibility was considered by Lewis that meltwater tunnels or caves beneath such snow patches, like bergschrunds behind glaciers, might provide access for cold air from outside. One such cave beneath a snow patch possibly 15 m thick was examined by Lewis (1939) and found to extend at least 50 m in from the snow margin. The outside air temperature at the time was $13°C$; the air in the cave stood at $9°C$, and meltwater in the cave less than $2°C$ above freezing-point. However, the very limited degree of air circulation in caves open only at one end makes it unlikely that freezing conditions in the cave would be re-established simply with each nocturnal fall of outside air temperature below zero. The fact that the base of the snow patch was still frozen in the daytime to the loose rock beneath in the drier inner parts of the cave does not necessarily imply that the cave acts as an avenue for penetration of cold air. The most important function of the cave is, rather, to act as a zone of thawing because of the concentration of meltwater along it.

The occurrence of freeze-thaw action in association with certain snow patches is firmly established. In all cases, however, the thickness of the snow, the presence or absence of permafrost beneath, and the frequency and amplitude of atmospheric freeze-thaw cycles must be taken into account. It will be apparent that the snow itself plays little part except in providing meltwater, and, if sufficiently thick, it may even hinder the process by protecting the ground from sudden temperature changes. The exact extent and significance of temperature changes beneath thick snow patches are difficult to assess, and so far have rarely been measured, for the digging of any artificial pit or

Table 5.1   Temperature observations, 8 July—7 August, in the vicinity of two snow patches in the Lake Louise area, Alberta (J. Gardner, 1969)

	Mean temperature °C	Mean minimum °C	Mean maximum °C	Number of ice days*	Number of frost alternation days†
Thermograph behind snow patch A	8·2	−0·1	16·5	1	17
Thermograph behind snow patch B	1·1	−1·4	3·6	18	5
Average values for two thermographs on rock wall: above A	15·8	6·6	25·0	0	0·5
above B	11·0	6·2	15·7	0	2
Data for Lake Louise (weather station)	13·2	3.1	23·3	0	0

* Temperature below zero throughout 24 hours
† Temperature crossed 0°C at least once in 24 hours

tunnel will obviously destroy the natural temperature régime that existed under completely closed conditions. But we may conclude that the most intensive zone of freeze-thaw action associated with any snow patch is peripheral to it and extends probably only a short distance inward beneath the thin edge of the snow.

J. Gardner (1969) has examined the temperature régimes associated with snow patches banked against steep rock walls at about 2300 m in the Lake Louise area of Alberta. Recording thermographs were placed both in the cleft behind the snow patches and higher up, above the snow patches, in cracks in the rock wall. Table 5.1 summarizes the results. The data refer to a summer period of one month up to 7 August, when snow patch A on a south-facing slope finally melted. Snow patch B lay on a north-facing slope and survived longer. Probable instrumental errors were in the range $\pm 1 \cdot 2°C$. The observations show that the mean, minimum and maximum temperatures are all reduced behind snow patches, and that there are more 'frost alternation days' as a result. Whereas air temperatures at Lake Louise and temperatures in cracks in the rock walls above the snow patches never fell below zero in the period of observation, sub-zero temperatures were recorded behind both snow patches. The temperature régimes approximated to Wiman's Icelandic type (p. 8), and although the mean rates of freezing were only about $0.5°C$/hour behind the snow patches, there were many signs that freeze-thaw action was encouraged in the vicinity of the snow: shattered bedrock was evident, and meltwater was seen to refreeze on making contact with the rockwall behind snow patch B.

Once the rock has been broken down by freeze-thaw action, its removal is essential if erosion forms are to develop. The main agencies available for transport of finely-broken debris beneath and downslope from snow patches are running water and gelifluction, and opinion has supported one or the other at times. W. E. Ekblaw (1918) favoured gelifluction moving material beneath the snow towards the lowest point; Lewis in his earlier paper on nivation (1936) refers to downhill creep and 'solifluction' facilitated by the generous supply of moisture beneath and near to the melting snow; T. T. Paterson (1951) also thought gelifluction more important than meltwater runoff in removing material from the base of snow patches in north-west Greenland. S. G. Boch (1946) observed that gelifluction terraces were frequently present below the lower edges of snow patches, and argued that the weight of the snow would itself encourage the process of gelifluction in the lubricated layer of debris beneath the snow patch (see Chapter 4). The evidence in favour of meltwater as a transporting agent is less impressive, though Lewis appealed to it in 1939. He noted how some 60–80 runnels emerging from beneath a snow patch on Snaefell (Iceland) were actively removing clay and sand in considerable quantity. McCabe in Spitsbergen was unable to produce such definite evidence in support of meltwater removal, but seems to have thought it the most probable process of transport. It may be that meltwater removal is dominant in relatively wet climatic conditions such as are characteristic of the part of Iceland in which Lewis worked, whereas Boch in the northern Urals and Paterson in north Greenland were concerned with nivation in colder drier conditions where gelifluction would prevail. In many cases, both processes probably operate.

Some other processes of weathering and denudation have at times been grouped under the heading of nivation, though their significance is probably slight compared with freeze-thaw action and the accompanying processes of debris removal. J. E. Williams (1949) suggested that chemical weathering might occur beneath snow-drifts. It is well known that the optimum temperature for the absorption of oxygen or carbon dioxide by water is near to $0°C$, and Williams showed that samples of air taken from inside a snow-drift contained more than twice as much CO_2 as normal. The probable reason for this is that, as water freezes, it liberates the CO_2 which, as a relatively heavy gas, then remains in the snow-drift ready to be absorbed again by meltwater. But the effectiveness as a weathering agent of water containing CO_2 even in abnormal concentrations is very much to be doubted, except where the bedrock is calcareous (M. Boyé, 1952). Moreover, rates of chemical reaction are reduced with lowering of temperature. Williams failed to consider any processes of chemical weathering other than those that may be connected with water containing O_2 or CO_2. A further weakness is that he failed to record the nature of the bedrock beneath the snow where his observations of CO_2 content were made.

Another process connected with the presence of snow patches is the possible erosive action of meltwater runoff. W. A. Rockie (1951) in the Palouse observed the development of rills and gullying by meltwater on the ground downslope from snow-banks, while in the very different environment of Antarctica, R. L. Nichols (1963) reported that meltwater dripping off the lower edges of snow patches on to the unprotected ground beneath was a significant eroding agent, as well as producing miniature gelifluction forms and fans of fine sand. On the other hand, Lewis (1939) thought the small trickles of water emerging from beneath snow patches in Iceland were not primarily eroding but transporting agents, as already mentioned. The function of the meltwater draining from a snow patch therefore varies according to the local conditions; there may be erosion if the water crosses unconsolidated material on a sufficiently steep slope, there may be mass movement if the debris is saturated, or the water may simply act as an agent of removal.

Although nivation does not include the effects of rapid snow movement in the form of avalanching, some writers have believed that slow mass movement of the snow is a prerequisite of surface erosion by nivation. Matthes (1900), in his original discussion of nivation, found no evidence of such movement, but other writers subsequently were not so sure. Thus, T. C. and R. T. Chamberlin (1911) stated, 'erosion is assumed to be due to the adhesion of the snow-ice mass to the ground on which it rests.... A broad patch of soil and loose rock ... is first dragged away ...' (p. 199), while I. Bowman (1916) contended.that 'in discussing the process of nivation it is necessary to assume a gliding movement on the part of the snow ...' (p. 286), and that 'compacted snow or névé of sufficient thickness and gradient may actually pluck rock outcrops ...' (p. 289). At the same time, Bowman denied that snow sliding would cause striation or abrasion. J. L. Dyson (1937), however, found clear evidence that rapid snow sliding could cause bedrock striations. Lewis (1939), in his studies of nivation hollows in eastern Iceland, did not consider that snow movement was a contributory process except that, in the case of one snow patch sloping at 15°, slow down-hill movement was seen to be taking place causing the formation of arcuate crevasses.

There is clearly much uncertainty about the occurrence and possible importance of slow mass movements of snow. The movements comprise both basal sliding and internal creep when conditions are appropriate, but there is disagreement as to whether such movements may significantly contribute to surface erosion. Recent investigations by A. B. Costin *et al.* (1964, 1973), have suggested that slow movements may lead to some surface abrasion, but not to an excessive degree. In association with a semi-permanent snow-patch in the eastern cirque of Mount Twynam, New South Wales, fresh abrasion marks on granodiorite bedrock were undoubtedly caused by recent movement of stones, some of which were still in position at the ends of their tracks. Both the stones and their tracks of movement sometimes bore traces of fresh rock flour. In this area, moving snow was thought to be the only possible agency capable of impelling the stones; moreover, avalanche action was out of the question for reasons such as the fact that some stones had come to rest in precarious positions and that avalanches are virtually unknown in the area. Slow mass sliding of the snow was therefore postulated. But although striation of bedrock surfaces was occurring, the resulting amount of erosion was probably quite small (and probably minor in comparison with other nivation processes). Patches of peat still adhered to the scratched bedrock surfaces, though it is possible that these peat masses would be protected by being sheathed with ice in the late autumn before the winter snowfall, and that the snow could then slip easily over them. Estimates of the range of shear stresses likely to be exerted on bedrock protrusions by the moving snow were obtained by inserting mild steel rods into drill holes in the bedrock. The rods were notched to give effective diameters of 0·25, 0·37, 0·55 and 0·79 cm and were placed in positions where snow-patch movement had previously been detected. Their break-points and bending resistances were first determined in the laboratory and their field behaviour recorded over 8 years. Those rods that failed recorded stresses from 1·6 to over 15·9 bar, forces ample to account for the observed signs of snow-patch erosion.

In areas of unconsolidated rocks, the rapidity with which nivation processes may work has often been remarked on. Following a severe snowfall on the Wasatch Plateau, Utah, in 1952, large snow-banks here persisted throughout the summer. In one season, the turf had been cut away to create nivation hollows one metre deep and up to 500 m across (R. F. Flint, 1957). Similar rates of action were recorded by W. A. Rockie (1951) in the Palouse of Washington State. In neither case was snow movement likely to have made any significant contribution to surface erosion.

On a longer time-scale, the nivation hollows studied by E. P. Henderson (1956) in Quebec province may have developed mainly in the Little Ice Age or at the most within the last 3000 years. The largest of them represent individually the removal of as much as 5 million m³ of material. The figures are very approximate, especially the time factor, but together with those of Flint and Rockie above, they suggest that nivation must be ranked as a competent and important process of denudation.

1.1   *Erosion features attributable to nivation*

The most important erosion feature is the nivation hollow. In south-east Iceland, 'the mountain slopes are everywhere scalloped by small hemispherical hollows irregularly spaced and showing all variations in size from a few hundred yards to over a mile across' (Lewis, 1936, p. 431). Lewis (1939) recognized three types of snow patch and corresponding nivation hollow: transverse, longitudinal, and circular. This classification by shape and position agrees well with and includes other workers' groupings.

Transverse nivation hollows lie along the contour of a slope and are frequently structurally determined. They include the hollows produced by Ekblaw's 'piedmont drifts' at the foot of steep lee slopes. Lewis notes that they may individually attain lengths of over a kilometre measured along the slope, and several hundred metres in a downslope direction; a downslope section through one is shown in Fig. 5.1A. This hollow fills up with snow completely in winter, and the snow may entirely disappear in summer. Vegetation ceases abruptly at the edges of the hollow. The ground at the lower part is almost flat, covered with stones and fine mud, across which meltwater trickles, while the upper part is steeper (up to 30°) and drier. It is suggested that the nivation processes already described result in the recession of the steeper upper slope into the hillside and the extension of the flat floor. The backslope thus grows in height, and the capacity of the hollow to hold snow is increased, until eventually enough snow is held to survive the summer melt period.

Transverse nivation hollows are the forerunners of cryoplanation terraces and cryopediments discussed in Chapter 6. In the area described by Lewis, the initial location of the hollows was determined by the sub-horizontally bedded basalts and tuffs, but slight depressions of non-structural origin might also be adequate to start the process, to which Matthes drew attention in 1900 (p. 182). H. W. Ahlmann (1919)

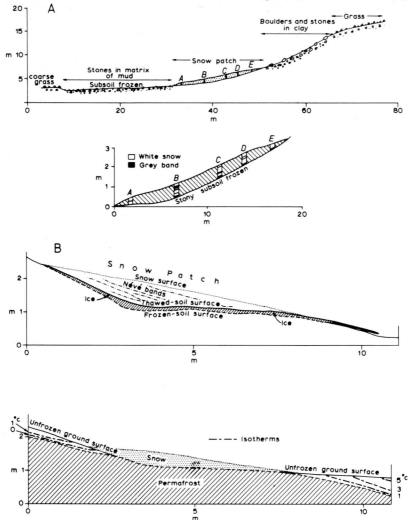

Fig. 5.1
A Downslope section through a transverse nivation hollow in Iceland (W. V. Lewis, *Geogrl J.*, 1939, Royal Geographical Society). The lower diagram gives details of the pits dug in the snow patch;
B Downslope sections through transverse nivation hollows in Spitsbergen (L. H. McCabe, *Geogrl J.*, 1939, Royal Geographical Society). On the lower section, isotherms (°C) for the unfrozen ground around the snow patch are shown. Beneath this snow patch permafrost rises to the ground surface.

also commented on the way in which snow patches might form and widen notches or shelves on mountain slopes (p. 229), and T. T. Paterson (1957) describes the long lines of snow-bank terraces picking out the strike of the Archaean sediments in northwestern Greenland. Similar forms, termed altiplanation terraces, are described by R. S. Waters (1962) in Spitsbergen.

In the Queen Elizabeth Islands, N.W.T., D. St-Onge (1969) describes how resistant gabbro is characterized chiefly by terrace forms, whereas outcrops of shale give rise to a greater variety of nivation features including amphitheatres, U-shaped widenings at gully heads and ledges. Not all terracing or notching of slopes caused by nivation is related to lithology or structure, as Gardner (1969) notes in the Lake Louise area.

A special case of miniature transverse nivation hollow is described by M. Boyé (1952) from the Pyrenees (Fig. 5.2). Snow lies at the foot of small limestone scarps in summer, covering them completely in winter. In summer, a gap appears between the head of a snow patch and its bounding scarp, in which frost-broken debris from the scarp accumulates, similar to a glacier randkluft. At the base of the scarp Boyé found a groove of semi-circular cross-section which he suggested marked the zone of most intense freeze-thaw action in the periglacial microclimate next to the snow patch; possibly also chemical weathering of the limestone by cold meltwater might play a part.

Longitudinal nivation hollows are those which follow the direction of maximum ground slope. The snow patches that give rise to them often accumulate initially

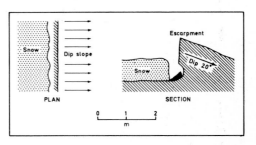

Fig. 5.2
Diagram to show the action of nivation undercutting a miniature scarp in limestone; central Pyrenees (M. Boyé, *Revue Géomorph. dyn.*, 1952)

in water-eroded valleys or gullies (Ekblaw thus termed these patches 'wedge drifts') and the major problem is that of separating the effects of water erosion from the possible effects of nivation (D. L. Dineley, 1954). In some it appears that nivation has widened the gullies, rounded their heads, and emphasized any structural ledges flanking the gullies. McCabe describes examples which possess a sharp cliff at their head over which a stream pours to plunge beneath the snow patch below the cliff. The streams, however, do not notch these cliffs and appear not to have formed them. In one particular case, the cliff overhangs, and the accumulation of angular boulders fallen down from the overhang into the gap between the rock wall and the snow patch could only have come from the overhang by freeze-thaw action. Water was seen to be trickling down the cliff and refreezing in the gap behind the snow patch.

Thus nivation modifies the original water-cut form. A later stage in the development of longitudinal nivation hollows in Vest-Spitsbergen is represented by the establishment of 'niche glaciers' (G. E. Groom, 1959). As the hollow is enlarged by nivation, so the snow drift can extend itself and thicken until ice begins to form by compaction in the lower layers. Such snow-ice masses form on some exceptionally steep slopes (42° noted by Groom); the characteristic landform is a rounded half-funnel shaped hollow. Pleistocene niche glaciers and 'proto-cirques' in Montana are described by Jacobs (1969).

Rather similar forms are the 'rasskars' or 'cirque-like ravines' on some precipitous rock slopes in Norway (Ahlmann, 1919). The head is funnel-shaped and the backwall may approach the vertical; downward, the funnel narrows into a V-shaped gorge. Running water is seldom seen in them. Their origin is not well understood, but snow collects in the funnel in winter, probably widens it by nivation together with frost action, and in spring the mass of snow and rock debris, lubricated by meltwater, glides down through the funnel outlet as an avalanche (see p. 155).

The third form of nivation hollow, related to roughly circular snow patches, differs basically from the transverse and longitudinal varieties in being largely independent of structural or water-eroded features. Varying from a few tens of metres in diameter to as much as 1 km, the rounded form is quite unrelated to stream action and is most perfectly developed on gently sloping surfaces where no pronounced variation in geological structure intervenes (R. J. Russell, 1933; Lewis, 1939). Fig. 5.1B shows a section through one measured by McCabe (1939), which should be compared with Lewis's section of a transverse nivation hollow (Fig. 5.1A). McCabe's section was measured in the month of July, before the snow patch dwindled away in August. In July, the ground around and away from the snow patch had thawed out to some depth, but beneath the snow patch, even though this was less than 0·6 m thick, the permafrost extended up to the ground surface, emphasizing the insulating effects of a snow cover

Plate XIII
Snow patch in circular nivation hollow, Devon Island, Canada. Vegetation stripes cross the pediment in the foreground. (C.E.)

mentioned earlier. Wherever permafrost is present at shallow depths, it has the extremely important effect of confining snow meltwater to the topmost layers of the ground. These areas are not only then most susceptible to frost action if refreezing later occurs, but also to gelifluction, a major process of debris removal. Away from the snow patches where the ground has thawed out to greater depths, what meltwater there is, is diffused throughout a considerable mass of material and its usefulness is greatly diminished. By the end of August when this particular snow patch had disappeared, the floor of the nivation hollow was seen to be covered with angular stones and coarse mud, and the ground was much drier than when the snow patch was present. Intensive freeze-thaw action was thus limited to the area of the snow patch and its

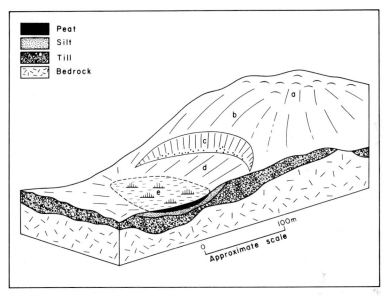

Fig. 5.3
Block diagram of a simple nivation hollow near Knob Lake, Quebec (E. P. Henderson, *J. Geol.*, 1956, University of Chicago Press); **a** numerous rocky outcrops on hill top; **b** slopes, 10–15°, in drift; **c** backwall, slope 30–40°, in drift, with boulder accumulations at its foot; **d** gently sloping (*c.* 5°) floor in drift; **e** swampy flat on gelifluction deposits

period of duration; and the rounded form of nivation hollow is contingent on the tendency of snow patches to become roughly circular by ablation (thus developing the least periphery per unit area) in any area where geological or existing topographical features do not play a dominating part.

Since evacuation of debris is necessary for the continued development of any nivation hollow, the circular type develops best on gently sloping rather than flat ground. The backslope forms an arcuate scar which gradually recedes, thus widening the floor and increasing the snow-holding capacity. In some areas of thick drift deposits near Knob Lake, Quebec, simple and compound nivation hollows of this sort are well portrayed (E. P. Henderson, 1956). A block diagram (Fig. 5.3) demonstrates the salient features. Backslopes, entirely in drift, reach heights of 20 m in places, with slopes usually of the order of 30–35°; the diameter may be 120 m for simple hollows or up to 800 m in the compound varieties where two or more hollows have coalesced or where mass movements have complicated the initial form. Floors slope outwards at 3–12° and may be covered with hummocky gelifluction material. The presence of thick unconsolidated and unresistant drift is the most important factor explaining both their distribution and the excellence of their development, for none occurs on rock slopes; and although some are orientated to face north-east, reflecting snow drifting with dominant south-westerly storms, the majority are orientated to face north or north-west in accord with the direction of last ice movement in the area. The direction of ice movement controlled the position and shape of the patches of thick drift in which the hollows are exclusively developed.

1.2   *Deposition features related to nivation*

Downslope from snow patches accumulate the materials removed by snow-melt run-off and by gelifluction. Gelifluction terraces may occur below snow patches (Boch, 1946) representing successive layers of geliflual material moving out from under the base of the snow and thickening as the gradient becomes less. The smaller terraces and layers may be seasonal. The scarps of some large terraces, however, may represent not the limits of gelifluction layers but the actual limits of the snow patch at various stages, as deduced by E. Watson (1966) in the case of a nivation hollow near Aberystwyth, Wales. Fig. 5.4 shows the series of terraces marking the exit of Cwm Du. The uppermost one is 20 m high and its face slopes at up to 23°. The terraces are built of gelifluction debris, which is thought to have moved down under the snow bank as Boch suggested. The evidence that the scarps mark successive snow-patch limits is two-fold: first, the present stream trench always cuts the scarps at their lowest points (where one would expect meltwater to leave a snow patch), and secondly, the scarps are not concentric but appear to enclose a series of lobes whose axes have progressively changed. This is more easily attributed to a snow patch whose headward growth was more rapid on one side than the other (in this case, the lee side, with respect to westerly winds).

   Boch claimed that a distinction might be drawn between the depositional effects of gently sloping and steeply inclined snow patches. In the former case, most debris broken off by freeze-thaw action would find its way to the base of the snow and be removed therefrom by gelifluction, thus producing terraces and platforms below the snow patch. In the case of steeply sloping snow patches, a major part of the debris falling on to the snow from the frost-shattered backwall would tend to slide down over the snow surface to build up a ridge at the foot of the snow patch. Such ridges were referred to by Russell (1933); C. H. Behre (1933) used the ambiguous term 'nivation ridges' and stressed that they were only incidentally connected with the presence of a snow bank since it was frost action above the level of the snow which produced the debris. The ridges were renamed 'protalus ramparts' by K. Bryan (1934). They have often been confused with moraines. R. P. Sharp (1942) describes examples on San Francisco mountain, Arizona; some ramparts here may nourish or even be the sources of stone streams since there is evidence that some ramparts are slowly moving downslope.

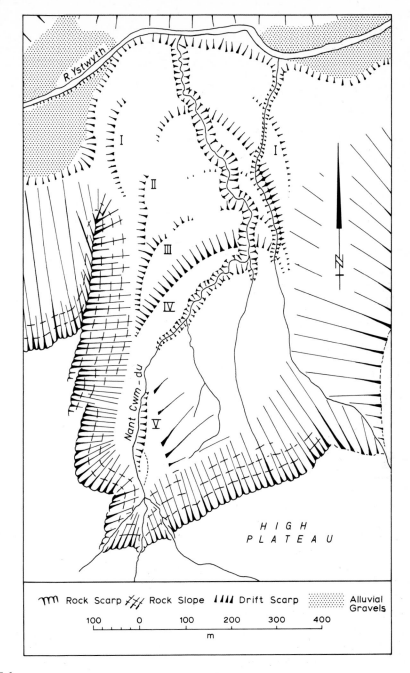

Fig. 5.4
Cwm Du, near Aberystwyth, central Wales, showing enclosing rock walls and a series of gelifluction fans I to IV. V is a protalus in the innermost recess (E. Watson, *Biul. Peryglac.*, 1966)

The distinction between gently sloping and steeply sloping snow patches is clearly an arbitrary one, but Watson's work at least suggests the limiting angle and supports the validity of the distinction. In Cwm Du where gelifluction deposits thickly mantle the floor of the hollow, the maximum snow slope is likely to have been less than 13°, whereas in neighbouring Cwm Tinwen, which possesses a protalus rampart, the snow surface may have been inclined at about twice this angle.

1.3   *The transition from snow patch to cirque glacier*

It has often been suggested—W. B. Wright (1914) was one of the first to do so—that nivation processes may enlarge and deepen initial depressions to such a degree that the thickening snow banks in them begin to acquire the characteristics of névé and eventually small glaciers. Theoretically, there should therefore be a graded series of landforms reflecting this transition, from the shallowest nivation hollow to the mature glacial cirque, but although isolated instances of transitional forms have been described, the field evidence is still inadequate. Some writers have used the term 'nivation cirque' (for example, Russell, 1933; Watson, 1966) to describe large nivation hollows resembling glacial cirques in form and situation but in which there may never have existed any true cirque glacier. Russell held that if striae and polished rock were found in such hollows, glacial action could be assumed, but since striae at least may be caused by moving snow rather than moving ice, the distinction on this basis is hardly valid. A much more fundamental distinction concerns the form of the floor of the hollow. A true glacial cirque will often possess a basin-shaped floor, with a reversed slope leading to the lip of the cirque; a nivation hollow formed by semi-stationary snow will always lack this characteristic, for removal of debris by gelifluction or other non-glacial means will not be compatible with any reversed floor slope. However, in practice, it may not be possible to demonstrate conclusively the presence or absence of a rock basin in the floor, owing to subsequent deposition, and rock basins may be absent in some cirques thought to be glacial.

The distinction between snow patch and glacier is no easier to make. Certainly, the presence of ice in the lower layers of a snow patch is not a satisfactory criterion, for the ice can form here by freezing of meltwater. Lewis (1939) argued that movement was the best distinguishing feature: 'when ... a composite mass of snow and ice begins to move over the rock floor, a new set of sculpturing agents is introduced' (p. 160). Yet slow mass movement of some snow patches is well established and there is no evidence that these are in the process of becoming miniature glaciers. Costin *et al.* (1964) thought that in the Mount Twynam area, the relationship was actually reversed, the processes of nivation (including snow sliding) following on from earlier glaciation and not leading up to glaciation.

Watson (1966) in west Wales attempted another approach by computing the probable maximum thicknesses of snow in 'nivation cirques' now vacated by snow. Assuming the snow surface at its maximum to have extended from the top of the backwall to either the protalus rampart (Cwm Tinwen) or the outer scarps of gelifluction terraces (Cwm Du), Watson shows that the greatest thickness of snow (measured vertically) was probably 38–45 m in the former case, and 75 m in the second. Since névé begins to form ice when subjected to a pressure of 30 m or more of overlying névé, some glacial ice was certainly present in Cwm Du (though perhaps for only short periods) but probably not to any extent in Cwm Tinwen.

The distinction between snow patch and glacier, or between nivation hollow and glacial cirque, cannot be rigidly formulated. Transitional forms do occur, though they have not yet been widely recognized or adequately investigated, and nivation may well be responsible for the initiation of some cirque glaciers.

1.4 *Stone pavements*

Flat or gently sloping pavements consisting of closely packed stones whose flat surfaces are uppermost appear to be related to persistent snow patches. In the mosaic of stones, smaller stones and particles occupy the cracks, and there is usually finer material beneath. Stone pavements have been noted on valley floors, near lakes or streams, and below breaks of slope. They have been described from Alaska (S. C. Porter, 1966), in Finnish Lapland (J. Piirola, 1969), in the Colorado Front Range (S. E. White, 1972) and in the Antarctic (A. L. Washburn, 1973). Earlier references by C. Troll (1944) are noted by White, but it is only quite recently that the distinctive characteristics and periglacial relationships of these pavements have been admitted. Formative processes are still unclear. White (1972) invokes frost-heave to raise the stones to the surface, after which the weight of accumulating snow helps to compact the material and embed the larger stones in a finer matrix, rotating and tipping them into a flat-lying position. He also points out that meltwater from the snow-bank will keep the ground sufficiently soft to permit movement of the stones (but the stones do not show any preferred orientation). Washburn (1973) adds suffosion (piping) as a possible process removing fines from the surface layer and notes that Aufeis as well as snow could be the cause of compaction. The diagnostic value of stone pavements as periglacial indicators is very uncertain and more study of these features is needed.

2 Nivation: summary and conclusion

The term nivation covers a variety of processes associated with a patchy snow cover, which give rise to the formation of distinctive erosional and depositional features. The typical erosion form is the nivation hollow which develops most rapidly in unconsolidated rocks such as glacial deposits, and which varies in form with the initial relief and geological structure of the region. Benching or terracing of hillslopes may also result from nivation. Depositional features include protalus ramparts at the foot of relatively steep snow slopes, and gelifluction terraces below more gently sloping snow patches.

The principal erosive process involved in nivation is freeze-thaw action. The effectiveness of this depends on rock type and the amplitude and frequency of atmospheric freeze-thaw cycles. The thickness of the snow cover is a most important consideration, as is also the existence of permafrost in the subsoil, for it is likely that thick snow patches on unfrozen ground play a largely protective role. In any case, the process is most effective and potent around the edges of the snow patch, and the action is extended areally by changes in the size of the snow patch. The snow itself acts mainly by supplying meltwater, which will be unable to soak away into the subsoil if permafrost is present. Chemical weathering beneath the snow is probably not very significant except in areas of calcareous rocks.

Debris resulting from freeze-thaw action must be removed from the site of the snow patch if erosion of the ground surface is to result. Removal can only be effected on sloping ground, apart from possible slight evacuation of material in solution by downward percolation in unfrozen ground. Gelifluction and meltwater runoff have both been considered as agents of transport, with the former appearing to be the more important in colder regions, and since neither can move material uphill, the floors of nivation hollows slope continuously outward.

Snow patches on ground of sufficient slope (yet not so steep as to cause avalanching) may move by slow basal sliding and/or internal creep. While this may give rise to scratched and abraded rock surfaces, it may not be a significant process of erosion until great thicknesses of névé have accumulated and the formation of glacier ice in the basal layers has begun. Thus snow patches may develop into small glaciers, and some nivation hollows may evolve into embryo cirques.

Since the snow itself has little direct part to play in nivation apart from providing meltwater and the negative aspect of insulating the ground surface from atmospheric temperature fluctuations, some workers have advocated abandoning the term nivation altogether. Paterson (1951) stated that 'the phenomenon of nivation ... is essentially a process concerned with the growth and melting of ice crystals within rock interstices, and as such has little to do with snow itself except in that the snow bank provides part of the water from which the ice crystals are developed' (p. 52), and M. Boyé (1952) asked: 'since snow neither attacks nor sweeps away, why speak of erosion by snow?' (p. 31). The validity of Boyé's criticism is somewhat doubtful in view of the evidence that slow-moving snow can in fact act as an agent of erosion, and even more so when moving rapidly as an avalanche. The term nivation remains a useful one to encompass a variety of processes connected with the presence of semi-stationary snow, provided that the limited part played directly by the snow is fully appreciated.

3 Avalanche activity

Avalanches are one of the most spectacular and dangerous forms of snow activity. Their geomorphological role has, however, not been very extensively studied. Nevertheless some valuable work has been done recently in this field, and features that are the direct result of avalanche activity have been recognized. In mountainous areas with heavy snowfall, avalanches are common, and it has been estimated that several tens of thousands take place in Switzerland each year. If the avalanches occur repeatedly in the same positions for many years in succession, their possible influence on the landscape cannot be ignored.

Snow is an extremely variable substance. E. La Chapelle (1970) lists seventy-nine varieties, of which ten are recognized by the International Snow Classification: plates, stellar crystals, columns, needles, spatial dendrites, capped columns, irregular particles, graupnel (soft hail), ice pellets and hail. As the snow becomes modified after falling

Plate XIV
Avalanche debris (snow and broken trees) in Yoho valley, Canadian Rockies. (C.E.)

to the ground, its high ratio of surface area to volume is reduced, and consequently the high surface free energy is lowered, so that the grain becomes rounded. Depth hoar forms from water vapour deep in the snow, especially with large temperature gradients, reducing the mechanical strength. This process occurs mainly in shallow snow and during very cold weather. Rime forms when supercooled water touches a surface, such as a hillside or ice. Rime-coated crystals bond more readily and do not deform, and thus can cause avalanche danger. Rime can also bond newly fallen snow, and prevent wind loss.

Avalanches can be of loose or slab form (see p. 149). Table 5.2 shows a classification of avalanches according to M. R. de Quervain (1966). Loose avalanches form when the wind is light or when metamorphism of previously cohering snow takes place. The snow is usually unstable so that large avalanches do not build up. Snow slabs are dangerous and occur when poorly bonded layers exist. Each snowfall produces a separate layer of distinct type and the bonds between the layers are weakened by depth hoar formation. Then tensile stresses can set off instability and avalanche activity.

Avalanches can be lubricated by an air layer, as discussed by R. L. Shreve (1968). The harmonic mean permeability must be less than 1 darcy, which is a reasonable value for poorly sorted debris. The product of permeability and bulk density of the basal debris must be less than 0·7 times the product of the harmonic mean permeability and the arithmetic mean density of the debris as a whole, which is probably true of landslides.

Not all avalanches significantly affect the land surface, but some types can produce particular landforms. Snow surface avalanches do not affect the ground. In their upper courses at least, the new snow slides above the plane separating it from the older snow beneath and it is not in contact with the ground beneath. An avalanche will take place

Table 5.2 Avalanche classification (M. R. de Quervain, 1966)

Criterion	Characteristics	
1 Type of rupture	*starting from a line:* slab avalanche (soft slab, hard slab)	*starting from a point:* loose snow avalanche
2 Position of sliding surface	*within snow cover:* surface layer avalanches (new snow fracture, old snow fracture)	*on the ground:* entire snow cover avalanche
3 State of humidity	*dry snow*	*wet snow*
4 Form of track	*open flat track:* unconfined avalanche	*channelled track:* channelled avalanche
5 Form of movement	*whirling through air:* airborne powder avalanche	*flowing along ground:* sliding avalanche flowing avalanche
6 Triggering action	*internal release:* spontaneous avalanche	*external trigger:* natural, artificial
7 Further important features: dimensions altitude quality of sliding layer time of descent velocity of descent		
8 Important genetic factors: amount and intensity of snowfall wind variations in snow and air temperature stratification of old snow deposits		

when the weight of snow exceeds the resistance of friction. The avalanche is affected by four factors:

1 the internal cohesion of the snow
2 the thickness and density of the snow layer
3 the character of the underlying material, whether it be old snow or rock
4 the slope.

The first factor affects the type of movement: the snow may either slip or break up into blocks. The second factor determines the mass of the avalanche. The third factor determines the frictional effect and the fourth determines the effect of gravity. There is a relationship between the slope and the thickness of the snow layer required to cause an avalanche. If the slope is 50°, an avalanche can theoretically occur with a layer only 5 cm thick; for a slope of 30°, the critical thickness is 15 cm, and for a slope of 22°, the thickness must increase to 40–50 cm before avalanching occurs. Normally avalanches do not occur on slopes less than 22° although exceptionally when all the factors are favourable they may occur on slopes as low as 6°.

Photogrammetry has been applied to avalanche studies in the Mount Temple area, Banff National Park (M. C. Van Wijk, 1967). Cameras were installed at both ends of a base line at the foot of a 45° slope. An avalanche was then released artificially and a map made from photographs taken before the avalanche, at intervals of 4·0, 11·2 and 18·4 s after its release, and also after it had come to rest. Contours were plotted at 1·5 m intervals to depict the rapidly changing form of the avalanche. From the data the speed of movement and volume were determined and the air turbulence studied.

The smoother the slope beneath the snow, the more readily an avalanche can occur. A dense vegetation cover, especially of forest, plays a large part in preventing avalanche activity. Once the vegetation has been removed and the rock smoothed by a series of avalanches or other processes, then avalanches can occur with greater ease. The avalanching process is, therefore, to a certain extent self-generating.

Avalanche activity depends not only on relief and surface characteristics but also on climate. Avalanches will be numerous and large where heavy snowfalls are common and temperature changes rapid, particularly those that cause the snow to melt or rain to fall on snow. The snow can absorb a large amount of rain or meltwater which increase its weight and avalanches then become much more probable. They also tend to occur more readily in areas where the tree-line is low for climatic reasons, because forests are effective in preventing avalanche activity. Statistics concerning avalanche activity in Switzerland show an interesting relationship between elevation and the number of avalanches. The tree-line in the area is about 1500 m, and the permanent snow level is about 3000 m. The number of avalanches recorded in 1909 was as follows:

under 1500 m	394	4 per cent
1500–2000 m	2632	28
2000–2500 m	3806	41
2500–3000 m	2210	24
over 3000 m	326	3

Most of the avalanches take place in the spring, when the snow cover reaches its maximum thickness and the temperature is liable to rise suddenly.

Observations of avalanches in France date back to 1899 but have only been systematically kept and recorded since 1959. In the 1960s, over 5000 avalanches were recorded as shown in Table 5.3. Precipitation is shown to be the main cause of avalanches, while higher winter temperatures are mainly a stabilizing factor. The wind is sometimes important, as is also the intensity, duration and type of snow fall. The general timing of avalanches can now be predicted but their precise location can not.

Avalanche frequency has been assessed by N. Potter (1969) using data from tree-ring analysis carried out on damaged trees. The evidence comprised datable scars on trees, changes of growth pattern of the rings due to tilting, or to destruction of adjacent

trees and measurements of the age of trees in reforested avalanche tracks. The first
and second types of evidence are the most reliable. Some tracks remain non-forested
in areas of frequent avalanches. Trees show scars above the low branches that were
protected by snow. The forested tracks have trees of different ages, in the form of trim
lines. Above the tree-line, avalanche boulder tongues (see p. 151) provide the best
evidence for persistent avalanche activity. Some erosion occurs on the avalanche
tongues as cleaner snow carries off upstanding boulders, which then form a zone around
the edge.

The location and timing of deep-slab avalanches in the Montana Rocky Mountains

Table 5.3   Avalanches in France, 1959–70 (J. J. Plas, 1970)

Year	Number of avalanches	Monthly distribution, all years, 1959–70	
1959–60	483		
1960–61	364	December	520
1961–62	298	January	1193
1962–63	593	February	1699
1963–64	101	March	1260
1964–65	439	April	720
1965–66	865	*Total*	5392
1966–67	270		
1967–68	578		
1968–69	435		
1969–70	966		
Total	5392		

depends on aspect according to C. C. Bradley (1970). Shaded and sheltered slopes tend
to avalanche in December, January, February and May. Exposed and sunny slopes
avalanche mainly in March and April. There is a cyclic rise and fall in the strength
of the basal snow layer, which on the whole becomes weaker as the winter progresses.
Avalanches in Savoy in the winter of 1969–70 have been described by C. Lovie (1970).
Snowfall was late, concentrated into two months, and mean temperatures were lower
than for the previous ten years. A long sequence of storms occurred, as a result of
which the areas of high and intermediate elevation had most avalanches in February
during cold, snowy weather, while the lower western area experienced most avalanches in March and April during the period of spring warming. The probability of
avalanche occurrence has been discussed by S. Takahashi *et al.* (1968). The fixed
elements in the situation are the vegetation, gradient, direction and type of slope.
With these variables, eighty-four classes of pattern were established from air photographs showing both snow-covered and snow-free areas. From these data, the statistical
probability and patterns of avalanching were assessed.

The Alaskan earthquake of 1964 set off an avalanche known as the Puget Sound
avalanche, described by M. C. Hoyer (1971). Puget Peak is 1198 m high; the slide
was nearly 3 km long and 0·5 km wide, moving 1,820,000 m³ of material which was
deposited as a blanket, up to 3 m thick in the cirque basin below the peak. Starting

as a rock fall, it reached the cirque travelling at more than 100 km/hour in places where it set snow in motion. Rock surfaces were stripped bare and rock scored. Shattered bedrock and steep slopes will cause recurrences. A post-avalanche mud flow occurred, in which rock debris was involved on the lower slopes. Many trees were destroyed and incorporated in the marginal deposits; sub-parallel ridges were formed in the cirque basin and on the slopes. The avalanche material travelled as far as the beach. As discussed by Shreve (1968), the avalanche probably moved on a cushion of compressed air: undisturbed snow remains below the debris in places. It took less than 3 minutes to move down the 3 km track; boulders up to 6·1 m were carried, the coarsest material being deposited on the beach.

3.1   *Types of avalanche*

The types of avalanche include powdery avalanches, slab avalanches and wet avalanches. The wet type of avalanche has been called 'slush avalanche' by some authors. There are also ice avalanches, falling from unstable and overhanging ice masses, sometimes sufficiently continuous in their activity to nourish a glacier below. All the main types of avalanche can be either clean or dirty. The clean avalanches consist only of clean snow or ice, while the dirty avalanches incorporate rock material in the moving mass. The latter will have the most important geomorphological effect. The geomorphological effects of these different types of avalanche have been studied quantitatively and in detail by A. Rapp (1959, 1960) in northern Sweden and by A. Jahn (1967) in Spitsbergen.

The effects of powdery avalanches are largely secondary. They are nearly always clean and consist of the movement downslope of newly fallen powdery snow of very low density. Their main effect is the result of their very rapid velocity of about 200–300 km/hour. This rapid movement creates a blast of air that can cause minor damage and trigger off other avalanches. At times they also create instability in the underlying snow, generating an avalanche that reaches to the ground and thus becomes a dirty avalanche.

The slab type of avalanche (see R. I. Perla and E. R. La Chapelle, 1970), which consists of large blocks of consolidated snow, is much more likely to become a ground avalanche and hence a dirty one. It can cause considerable erosion. The blocks of snow become rounded as they move down the slope and in this form they can move far from the mountain slope. They can carry debris a considerable distance, even extending a short way up the opposite valley slope. They may dam up small streams in the valley, and floods occur when the wet snow dam suddenly gives way. Their effect is, therefore, both direct and indirect.

The wet or slush avalanches have probably the most important direct geomorphological effects. It has been reckoned that this type of avalanche can carry about ten times as much material, under certain circumstances, as the other types, according to Rapp's measurements in northern Sweden.

The effects of slab avalanches in Kärkevagge have been considered by Rapp (1959). He describes one avalanche that was set off when a block of snow 50 m long and 2–3 m thick fell from the top of a cirque wall. The snow was hard and heavy and broke into pieces as it fell down the very steep slope, which reached 90° at the top. The snow fell on to rather less steep slopes of 50–60° and continued down the slope as an avalanche. The falling snow detached several large blocks of rock about 0·5–1 m in size. One of these types of avalanche, which fell in July, transported about 5–10 m³ of rock debris. This avalanche consisted of old firn snow and removed loose debris liberated by frost action from the rock wall. These relatively small avalanches of hard snow provide a means whereby rock surfaces can be cleaned of superficial debris and the avalanche can at times break off rocks loosened by frost.

The depositional forms resulting from this type of avalanche consist of debris that is roughly sorted according to size. The larger rocks roll farther down the slope than the smaller ones, but otherwise the debris is completely mixed, fine material being mingled with coarse. The form of the deposits is usually in the shape of a broad tongue. The smaller avalanches do not extend far down the slope, but the larger ones sometimes reach the valley floor or beyond. Some of the boulders that come to rest on or in the avalanche material move farther down the slope as they melt out. Avalanche debris can often be recognized by its loose packing as it melts out of the snow, and by the spread of finer material on larger boulders that were originally at the bottom of the avalanche. The boulders and cobbles in the melted-out avalanche debris show no preferred orientation, although the elongated boulders that have slid over the snow are often orientated with their long axes downslope.

The moving boulders often cause some erosion of the ground over which they move. Examples of this type of erosion were seen in Austerdal, Norway. Boulders carried by avalanches had moved very fast down the steep valley walls and had been broken up on impact with boulders already lying on the valley floors. These boulders were also broken by the impact into many angular fragments. The dirty snow avalanches occur mainly where deep snow drifts can accumulate during the winter. They tend to recur in the same places year after year, and are particularly frequent on lee slopes. They occur mainly in the spring as the snow melts. Dirty avalanches are common in late spring when the thaw has exposed much bare ground. Rapp estimated the total amount of material moved annually by dirty snow avalanches to be of the order of 1·4 t/m², moving an average distance of 100 m on a 30° slope. This gives a total of 1050 t/m vertically.

Slush avalanches have been found to be much more effective agents of denudation than dirty snow avalanches. They are particularly important in years of rapid snow melt. The operation of slush avalanches or slushers has been described in Greenland and arctic North America. Slush avalanches are sometimes initiated by the melting of snow in a stream bed. Water may be held up behind an ice dam for some time, and then suddenly escapes down the valley when the dam gives way. An example of this kind of slush avalanche occurred on 12 June 1956 at Kärkevagge. This avalanche carried 200 m³ of rock material and extended for 350 m, reaching gently sloping ground. A still larger one occurred in 1958, carrying 300 m³ of debris. These avalanches were released from an iced-up waterfall and as they fell, strips of turf were torn from the ground, exposing the bedrock. The carrying capacity of such avalanches is indicated by the fact that a boulder 5 × 3 × 2 m was carried about 120 m down a slope of only 5°. The boulders plucked from the ground first glide over the ground and then they may move by jumping; they may then form hollows where they land. Finally they roll and make smaller holes, closer together. Sometimes the holes from which the boulders were plucked can also be seen. Slush avalanches can be released on gentler slopes than normal avalanches. The saturation of the snow, which increases in weight as a result, is partly the cause. Layers of ice on the lower side of the snow patch may provide glide planes to initiate the avalanche. Sometimes slush avalanches start as a snow slab avalanche that moves into a stream valley. The snow then mixes with the water and is converted into a slush avalanche. Three major slush avalanches occurred during the period from 1952 to 1960 in this part of northern Sweden, and it is estimated that these three carried a total of at least 700 m³ of rock debris.

There seems to be a difference between the slush avalanches of the arctic regions and the dirty ground avalanches of the Alps. Some of the particularly disastrous Alpine avalanches may, however, have had the characteristics of slush avalanches. The Alpine ones differ from the arctic type in that the Alpine ones are released as large slab avalanches on steep slopes. They do not originate directly in stream courses. They tend to move down steep ravines and the material they carry is deposited on alluvial cones. Relief factors, therefore, are probably responsible for the major differences between the Alpine and arctic types of avalanche. In the Alps, there are steep incised gullies, but in the arctic areas, more open slopes are typical. N. Caine (1969) has noted that slush avalanches feed the lower slopes more effectively, both by redistribution of material and by the introduction of new material. Together these processes produce a concave slope profile. A model of slope development by slush avalanching is proposed, based on correlations between amount of debris accumulation and distance downslope.

3.2 Avalanche boulder tongues

The avalanche boulder tongues described by Rapp (1959) in Lapland are a good example of a feature that can be specifically related to avalanche activity. These constructional features also provide evidence of the capacity of avalanches to erode the slopes across which they are moving. The avalanche boulder tongue must first be differentiated from other features resulting from mass movement and stream flow, including talus cones and alluvial cones. Another rather similar form is that produced by a rockslide tongue. The rockslide tongue, however, lacks the long, straight path leading down to it which is characteristic of the avalanche boulder tongue. The normal avalanche boulder tongue also has a markedly concave longitudinal profile. The end of the tongue may reach far out into the valley or even extend a short distance up the opposite valley slope. The surface of the tongue is flattened and in this it differs from the talus and alluvial cones, which have convex surfaces.

There are also smaller features characteristic of avalanche boulder tongues, such as avalanche debris tails. The tails consist of small straight ridges of debris on the distal sides of large boulders. They are about the same height and width as the boulders, which may attain dimensions of 1 m, behind which they have formed. The tail is often about 5–10 m long, diminishing in size towards the distal end. All the tails are parallel to each other and are elongated in the direction of the tongue. There is also sometimes a small accumulation on the proximal side of the boulder. The features form in the same way as a sand shadow round an obstacle, or, on a larger scale, crag and tail. They are an accumulation and not an erosional form.

The whole boulder tongue probably takes quite a long time to form and is not the result of one single avalanche. Two types of avalanche boulder tongues are distinguished: road-bank tongues and fan tongues. Most of both types are asymmetrical in outline transverse to their elongation. The tongues consist of angular debris that is only very poorly sorted to the extent that the larger boulders tend to occur round the margin. The features form most effectively above the tree-line and in those areas where much loose rocky sediment is available on the hillsides.

One example of the road-bank type of avalanche boulder tongue was 330 m long from the mouth of the chute and 70–80 m broad, ending in a lake. It attained a thickness of 5 m at a point 150 m from its end. The material consisted of angular boulders up to 2 m in size. The long profile decreased from 28° to 22° and finally flattened to 15° near its end. The asymmetrical transverse profile is thought to be associated with snow banks that collect alongside the tongue. The snow becomes thicker on the lee side of the tongue, while the windward side is swept bare. Snow will accumulate in the chute above the tongue and hence promote avalanche activity. When the avalanche runs down the chute, it will tend to erode the exposed part. The material will accumulate around the bare part of the fan as the snow-bank on the lee side will tend to divert the avalanche to the bare windward side. In this way, the flattish top of the feature and its asymmetry can be accounted for. In transverse profile the lee side is steeper as the material slumps when the snow-bank melts.

The fan type of avalanche boulder tongue sometimes extends on to the flat ground in the valley floor. The fan starts in a broad shallow avalanche track about 200 m wide, wider than the chutes that lead down to the road-bank type. One fan-type avalanche boulder tongue was 650 m long and sloped at 35° on the lower part of the mountain wall. The slope decreased to 30° above the wall, a slope that is less steep than the angle of rest of boulders which cover the hillside where no avalanche has removed them. The upper part of the tongue consisted of an alluvial cone. This cone has been built up by streams and its eroded margins indicate that avalanches scour a wider area than that eroded by the streams. The true avalanche tongue reaches a further 300 m beyond the bottom of the alluvial fan. This indicates that avalanches carry material much farther than streams or rock-falls. At its lower limit, the tongue rises about 0·5 m above the grass surrounding it, but its general thickness appears to be about 2 m. The extent of individual avalanches is indicated by arcs of boulders on the tongue. A study of the preferred orientation of the stones on the surface of the boulder tongue showed that they are nearly all aligned parallel to the elongation of the tongue, but they show a slight divergence towards the lateral margins (Fig. 5.5). The fan type of boulder tongue is probably the work of very rare large avalanches that have the capacity to erode and carry large quantities of debris. It seems probable that this type of feature only occurs above the tree-line. They are widespread under suitable conditions in the northern parts of Scandinavia, where aerial photographs show that the hillsides are furrowed with the parallel features formed as a result of avalanche activity. W. E. Yeend (1972) considers that some unusual mounds of un-

sorted debris in the Brooks Range, Alaska, could have originated in this way. The
mounds were probably carried along chutes on snow and deposited in hollows in the
snow and at its edge. They cover 4–8 ha, and range in height from 0·6 to 7·6 m, conical
in form. They occur on the valley floor, as far as 300 m from the steep valley side.
He suggests the term 'winter protalus mounds' to describe them.

So far the depositional aspects of avalanche activity have been stressed, but it is important to recognize that the deposition must be accompanied by an equal erosion

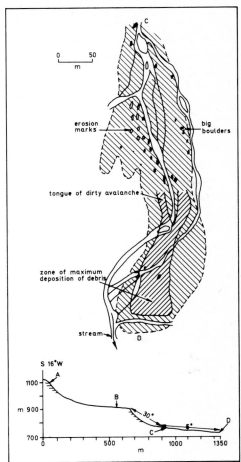

Fig. 5.5
An avalanche boulder tongue in Kärkevagge, Sweden (A. Rapp, *Geogr. Annlr,*
1956)

of material from the upper slopes. This activity influences the slope development processes in the areas where avalanche activity is a significant morphological agent. The
road-bank tongues lead up to narrow, fairly deeply cut chutes, which are very straight.
These chutes are also used by rock-falls and streams. Avalanches are, however, the
most efficient mechanism whereby rock and debris are carried down these chutes. The
avalanche removes loose material lying on the surface across which it is moving as
well as abrading this surface. Weathering is thus allowed to proceed effectively. Not
all rock types are equally susceptible to avalanche activity. Amphibolite appears to

react more obviously to avalanche erosion than the mica-schist of Kärkevagge, where avalanche chutes are not common.

The effects of avalanches in the Bundervista valley of Bulgaria, where there are large accumulation forms, have been studied by C. D. Peev (1966). Avalanche erosion takes place in the snow catchment area and throughout the avalanche tracks. Nivation is active in the hollows and freeze-thaw takes place in summer, giving debris for avalanche removal in winter and providing water in summer. Avalanches can completely clear their tracks of loose debris because of the steep slopes and the speed at which they travel. Loose rocks that are carried down erode the gully walls and break off further fragments. In the Bundervista valley the avalanches carry much rock and cross the valley to mount the opposite slope. A lake 14 m deep was partially filled, reducing the depth to only 6 m after one avalanche. Rock cones 5–6 m high form in angular, unstable material, while mounds form on the opposite side of the valley. Pseudo-moraines develop beside the track at the foot of the slope, and parallel ridges of unsorted material, looking like snow-plough ridges several metres high, form on flat terrain at the bottom of the valley. Larger avalanches erode the cones of smaller ones. Avalanche pits up to 300 m² in size form where impact is greatest. Avalanche snow cones melt only slowly and may survive the summer.

J. Gardner (1970) has considered the geomorphological significance of avalanches in the Lake Louise area of Alberta. They are most effective as geomorphological agents in areas devoid of vegetation. Avalanche boulder tongues of the fan and road-bank types occur, together with avalanche debris tails. In the area studied, half the mean annual precipitation falls as snow, which lies 1–3 m deep in the valleys for up to six months of the year. The mountains are high and steep. In winter from November to March, both dry and wet snow, direct and delayed avalanches occur, while from April to June, wet snow thaw avalanches occur. Early July has the modal value of 4·7 avalanches per hour of observation. Most occur between 1200 and 1600 hr, with the mode at 1300 hr. Most avalanches occurred above 2500 m and 90 per cent on shaded leeward slopes facing between north and south-east. Avalanches are only one of several slope processes, including rock-falls, debris and mudflows, and icefalls. A large spring thaw avalanche of 45,000 m³ of snow was seen on an avalanche boulder tongue of road-bank type, with boulder content estimated at 14·3 m³, some of the boulders being more than 1 m³. Avalanche boulder tongues are the most obvious feature and are mostly of the fan type. Road-bank types occur but are rare, although a continuum exists. The fan and road-bank types are concave in long profile (Fig. 5.6), the range of slope angles being 9·5–33° on the fan type and 5·8–13·° on the road-bank type. Debris tails 5–10 m long and 0·5 m high occur on slopes of 10–30°. Precariously balanced boulders can result from the melting of snow in an avalanche. Winter avalanches tend to move over a snow surface and are, therefore, less significant geomorphologically.

The effectiveness of avalanches in modifying the slopes depends partly on the slope gradient. The optimum slope appears to be about 30–35°. On steeper slopes, although some modification does occur, it is not so marked. On slopes gentler than 30–35° rock-falls lose their effectiveness. One of the most important functions of avalanche denudation is to keep the slope bare of rock debris so that weathering can continue unimpeded.

Where the slope is swept bare by avalanches it will tend to retreat parallel to itself, with a growing concavity at its base. A talus slope will not replace the rock slope. The fact that avalanches erode wide tracks means that the valley walls will tend to remain fairly smooth and undissected, apart from the relatively shallow chute formation. In areas such as Lapland, avalanche activity could be responsible for the asymmetry of the valley walls. The east-facing walls are steeper than the west-facing ones. This variation is partly structural, but also in part it is the result of avalanches steepening or maintaining the steepness of the east-facing slopes. On the lee side of the hills, where

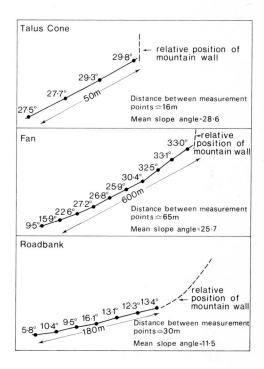

Fig. 5.6
Some representative slope profiles for talus cones, roadbank and fan slopes developed by avalanching (J. Gardner, *Arctic and Alpine Research*, 1970). The slope angle at each point is the mean of ten random measurements.

large snow-drifts can accumulate, and on cols, where the relief is suitable, avalanche activity will be great. The slopes in the lee of the cols and hillsides will tend to retreat by avalanche erosion. The total effect of avalanche erosion in the denudation of a steep hillside is difficult to assess. But it seems likely that this aspect of slope formation, particularly in mountainous districts with heavy snowfall, such as the Alps, has been under-estimated in the past. Avalanches in the Alps are partly responsible for the steep narrow chutes that are also used by summer storm waters and meltwater. These gullies are probably kept clear mainly by avalanche activity in the winter and spring. The avalanche chutes in Norway have been called 'rasskars' by Ahlmann (1919) and they are characteristic features of many hillsides.

Avalanches, therefore, produce characteristic features of both deposition and erosion. They can play an important part in the modification of the landscape in mountainous areas where there are heavy falls of snow and rapid summer melting.

4 References

AHLMANN, H. W. (1919), 'Geomorphological studies in Norway', *Geogr. Annlr* **1**, 1–148 and 193–252

ALLIX, A. (1924), 'Avalanches', *Geogrl Rev.* **14**, 519–60

(1954), *L'action morphologique de la glace et celle des coulées de neige* (Mél Bénévent, Gap), 11–17

BEHRE, C. H. (1933), 'Talus behaviour above timber in the Rocky Mountains'. *J. Geol.* **41**, 622–35

BOCH, S. G. (1946), 'Les névés et l'érosion par la neige dans la partie nord de l'Oural', *Bull. Soc. géogr. U.S.S.R.* **78**, 207–34 (original in Russian; translated by C.E.D.P., Paris)

BOWMAN, I. (1916), *The Andes of Southern Peru* (*Am. Geogr. Soc. Spec. Publ.* **1**)

BOYÉ, M. (1952), 'Névés et érosion glaciaire', *Revue Géomorph. dyn.* **3**, 20–36

BRADLEY, C. C. (1970), 'The location and timing of deep slab avalanches', *J. Glaciol.* **9**, 252–61

BRYAN, K. (1934), 'Geomorphic processes at high altitudes', *Geogrl Rev.* **24**, 655–6

BULL, A. J. (1940), 'Cold conditions and landforms in the South Downs', *Proc. Geol. Ass.* **51**, 63–70

CAINE, N. (1969), 'A model for alpine talus slope development by slush avalanching', *J. Geol.* **77**, 92–100

CHAMBERLIN, T. C. and CHAMBERLIN, R. T. (1911), 'Certain phases of glacial erosion', *J. Geol.* **19**, 193–216

COOK, F. A. (1962), 'Simple transverse nivation hollows at Resolute, N.W.T.', *Geogr. Bull.* **18**, 79–85

CORBEL, J. (1958), 'Climats et morphologie dans la Cordillière Canadienne', *Revue can. Géogr.* **12**, 15–48

COSTIN, A. B., JENNINGS, J. N., BAUTOVICH, B. C. and WIMBUSH, D. J. (1973), 'Forces developed by snowpatch action, Mt. Twynam, Snowy Mountains, Australia', *Arct. alp. Res.* **5**, 121–6

COSTIN, A. B., JENNINGS, J. N., BLACK, H. P. and THOM, B. G. (1964), 'Snow action on Mount Twynam, Snowy Mountains, Australia', *J. Glaciol.* **5**, 219–28

DAVEAU, S. (1958), 'Cône central d'éboulis de l'aiguille Rousse', *Revue Géogr. alp.* **46**, 423–8

DETTERMANN, R., BOWSHER, A. and DUTRO, T. (1958), 'Glaciation on the Arctic slope of the Brooks Range', *Arctic* **11**, 43–61

DINELEY, D. L. (1954), 'Investigations in Vest-Spitsbergen', *J. Glaciol.* **2**, 379–83

DYSON, J. L. (1937), 'Snowslide striations', *J. Geol.* **45**, 549–57

EKBLAW, W. E. (1918), 'The importance of nivation as an erosive factor, and of soil flow as a transporting agency, in northern Greenland', *Proc. natn. Acad. Sci. U.S.A.* **4**, 288–93

FLINT, R. F. (1957), *Glacial and Pleistocene geology*

GARDNER, J. (1969), 'Snowpatches: their influence on mountain wall temperatures and the geomorphic implications', *Geogr. Annlr* **51**A, 114–20

(1970), 'Geomorphic significance of avalanches in the Lake Louise area, Alberta, Canada', *Arct. alp. Res.* **2**, 135–44

GROOM, G. E. (1959), 'Niche glaciers in Bünsow Land, Vest-Spitsbergen', *J. Glaciol.* **3**, 369–76

HENDERSON, E. P. (1956), 'Large nivation hollows near Knob Lake, Quebec', *J. Geol.* **64**, 607–16

HOYER, M. C. (1971), 'Puget Peak avalanche, Alaska', *Bull. geol. Soc. Am.* **82**, 1267–84

JACOBS, A. M. (1969), 'Pleistocene niche glaciers and proto-cirques, Cataract Creek valley, Tobacco Root mountains, Montana', *Geol. Soc. Am. Spec. Pap.* **123**, 103–14

JAHN, A. (1967), 'Some features of mass movement on Spitsbergen slopes', *Geogr. Annlr* **49**, 213–25

LA CHAPELLE, E. R. (1970), 'From snowflake to avalanche', *Nat. Hist.* **79**, 30-9

LEWIS, W. V. (1936), 'Nivation, river grading and shoreline development in south-east Iceland', *Geogrl J.* **88**, 431–47

(1939), 'Snow-patch erosion in Iceland', *Geogrl J.* **94**, 153–61

LOVIE, C. (1971), 'Les avalanches de neige en Savoie durant l'hiver 1969–70', *Annls Met.* N.F. **5**, 235–9

MATTHES, F. E. (1900), 'Glacial sculpture of the Bighorn Mountains, Wyoming', *U.S. geol. Surv. 21st A. Rep.* (1899–1900), 167–90

(1938), 'Avalanche sculpture in the Sierra Nevada of California', *Bull. Un. géod. géophys. int.* **23**

McCABE, L. H. (1939), 'Nivation and corrie erosion in West Spitsbergen', *Geogrl J.* **94**, 447–65

NICHOLS, R. L. (1963), 'Miniature nivation cirques near Marble Point, McMurdo Sound, Antarctica', *J. Glaciol.* **4**, 477–9

PATERSON, T. T. (1951), 'Physiographic studies in north-west Greenland', *Meddr Grønland* **151**, 4, 60 pp.

PEEV, C. D. (1966), 'Geomorphic activity of snow avalanches', *Proc. int. Symp. (Scientific aspects of snow and ice avalanches)*, *Int. Ass. scient. Hydrol. Publ.* **69**, 357–68

PERLA, R. I. and LA CHAPELLE, E. R. (1970), 'A theory of snow slab failure', *J. geophys. Res.* **75**, 7619–27

PIIROLA, J. (1969), 'Frost-sorted block concentrations in western Inari, Finnish Lapland', *Fennia* **99**(2), 35 pp.

PLAS, J. J. (1971), 'Les avalanches dans les alpes françaises, dix ans d'observations et de mesures', *Annls Met.* N.F. **5**, 241–6

PORTER, S. C. (1966), 'Pleistocene geology of Anaktuvuk Pass, Central Brooks Range, Alaska', *Arct. Inst. N. Am. Tech. Pap.* **18**, 100 pp.

POTTER, N. (1969), 'Tree ring dating of snow avalanche tracks and the geomorphic activity of avalanches, north Absaroka mountains, Wyoming', in *U.S. Contributions to Quaternary Research*, INQUA VIII (ed. S. A. SCHUMM and W. C. BRADLEY), *Geol. Soc. Am. Spec. Pap.* **123**, 141–65

QUERVAIN, M. R. DE (1966), 'On avalanche classification—a further contribution',

Proc. int. Symp. (Scientific aspects of snow and ice avalanches), Int. Ass. scient. Hydrol. Publ. **69**, 410–17

RAPP, A. (1959), 'Avalanche boulder tongues in Lapland, a description of little known forms of periglacial accumulations', *Geogr. Annlr* **41**, 34–48

—— (1960), 'Recent development of mountain slopes in Kärkevagge and surroundings, north Scandinavia', *Geogr. Annlr* **42**, 65–200 (esp. 122–47)

ROCKIE, W. A. (1951), 'Snowdrift erosion in the Palouse', *Geogrl Rev.* **41**, 457–63

RUSSELL, R. J. (1933), 'Alpine landforms of western United States', *Bull. geol. Soc. Am.* **44**, 927–49

ST.-ONGE, D. (1969), 'Nivation landforms', *Geol. Surv. Can. Pap.* **69–30**, 12 pp.

SHARP, R. P. (1942), 'Multiple Pleistocene glaciation on San Francisco Mountain, Arizona', *J. Geol.* **50**, 481–503

SHREVE, R. L. (1968), 'Leakage and fluidization in air-layer lubricated avalanches', *Bull. geol. Soc. Am.* **79**, 653–8

TAKAHASHI, S., KIMATA, K. *et al.* (1968), 'The probability of occurrence of the avalanche', *J. Japan Soc. Photogramm.* **7**, 117–24

TRICART, J. and CAILLEUX, A. (1962), *Le modelé glaciaire et nival* (Paris), 198–207

TROLL, C. (1944), 'Strukturböden, Solifluktion, und Frostklimate der Erde', *Geol. Rdsch.* **34**, 545–694 (English translation, *Snow Ice Permafrost Res. Establ.*)

WASHBURN, A. L. (1973), *Periglacial processes and environments*

WATERS, R. S. (1962), 'Altiplanation terraces and slope development in Vest-Spitsbergen and south-west England', *Biul. Peryglac.* **11**, 89–101

WATSON, E. (1966), 'Two nivation cirques near Aberystwyth, Wales', *Biul. Peryglac.* **15**, 79–101 ·

WHITE, S. E. (1972), 'Alpine sub-nival boulder pavements in Colorado Front Range', *Bull. geol. Soc. Am.* **83**, 195–200

VAN WIJK, M. C. (1967), 'Photogrammetry applied to avalanche studies', *J. Glaciol.* **6**, 917–33

WILLIAMS, J. E. (1949), 'Chemical weathering at low temperatures', *Geogrl Rev.* **39**, 129–35

WRIGHT, W. B. (1914), *The Quaternary Ice Age* (London, 1st Ed.), 6

YEEND, W. E. (1972), 'Winter protalus mounds, Brooks Range, Alaska', *Arct. alp. Res.* **4**, 85–7

6

Cryoplanation, tors, blockfields and blockstreams

At the higher elevations in middle and high latitudes, a range of features reflects the very considerable activity of periglacial processes, particularly frost action, gelifluction and nivation. The formation and even the nature of some of these features are far from being clearly understood, and in some cases a periglacial origin is contested. Equivalent features of Pleistocene age have been recognized in areas where, today, the climate is not sufficiently severe to allow their development.

1 Cryoplanation

The term cryoplanation was introduced by K. Bryan (1946) to describe the combined effect of degradation by frost and the subsequent removal of waste by downslope movement, including gelifluction, and by running water and wind. The term is preferable to, but identical in meaning to, the older term 'altiplanation', which was devised by H. M. Eakin (1916) to signify a range of cold-climate processes and their effect in producing a distinctive assemblage of landforms. In the part of Alaska which Eakin studied, these forms comprised flattened summits, bench-like features occurring irregularly on spurs and hillsides, and bounding scarps of bedrock or of bare angular talus lying steeply at its angle of repose and including blocks up to 4 m in size. Summit flats and hillside benches admit of many explanations entirely unconnected with periglacial conditions; Eakin was the first to draw attention to the possibility that in some areas, these features might be produced simultaneously at a variety of altitudes irrespective of such considerations as base-level, and that their levels might have no significance in the denudation chronology of a region.

Eakin's account of the formation of altiplanation terraces is confused but it is evident that he considered the terraces mainly as constructional features, built out of rock waste derived by frost from the local bedrock; movement of the rock waste occurred by upward frost-heaving and slow transfer to the outer edge of a terrace. On the inner parts of a terrace, comminution of debris by frost was active since here conditions are likely to be wettest, so that gradually all the interstices of the fractured rock debris are filled with fines to give an even surface appearance.

Similar terrace features were described a few years previously by L. M. Prindle

(1905) and by D. D. Cairnes (1912), though neither used the term altiplanation. Cairnes noted that the fronts of some terraces consisted of solid rock but that behind the scarp, the gently sloping terrace surface was one of accumulated debris.

In the USSR, studies of cryoplanation features have been in progress even longer. One of the earliest references to cryoplanation terraces is by N. M. Kozmin (1890); the term 'goletz' terrace has been commonly used, and their occurrence noted in such areas as the northern Urals and Siberia. A useful review of the early work is given by G. Jorré (1933), who summarizes the main features of the terraces thus: downslope extent up to several hundred metres, vertical spacing usually 5–7 m, scarp slope of 20–50°, a gentle terrace surface sloping at 1–12° outward to the scarp edge, a marked lack of relationship with bedrock structure, and a debris cover over bedrock usually less than 3 m thick. J. Demek (1969a) notes exceptional examples attaining lengths of 10 km and widths of 3 km. Above the highest platforms rise 'stone cities' or huge tor-like rocky outcrops. The terraces do not occur below the lowest Pleistocene snow-line. The major part of their formation is attributed to frost, breaking up the bedrock and comminuting debris to silt-size particles, while snow-melt runoff is held to be responsible for transport. The surfaces of the terraces are often decorated with smaller cryogenic forms—sorted and non-sorted polygons, stripes, frost cracks—and

Plate XV
Cryoplanation terrace, Devon Island, Canada. The frost-riven bluff behind is partly covered by the snow bank. (C.E.)

forms resulting from downslope movement such as garlands, turf-banked terraces, gelifluction tongues and stone streams.

The first to recognize cryoplanation terraces in Britain was A. Guilcher (1950); he
described them on Holdstone Down, Great Hangman and Trentishoe Down in north
Devon. Significantly, rock exposures showed that at least one such bench was cut in
bedrock; in other cases, a thin debris cover was likely. Nivation played a part in the
initiation of the terraces; patterned ground developed on their surfaces, and breakdown
of material by frost and chemical weathering facilitated its downslope movement.
Guilcher's work was extended by Te Punga (1956), and it was confirmed that the terraces were essentially cut in bedrock, with a waste mantle no more than a metre or
so thick. Te Punga suggested that the terraces represented the work initially of transverse snow patches (Chapter 5); subsequently they were enlarged by frost action and
meltwater removed some debris, so that the treads were slowly lowered and the bounding scarps retreated.

A more recent and systematic study of cryoplanation terraces in the Dartmoor area
of south-west England has been undertaken by R. S. Waters (1962). The following
are the main features of the terrace morphology:

Height of scarps	=	2–12 m
Dimensions parallel to regional slope	=	10–90 m
Dimensions transverse to regional slope	=	up to 800 m
Slope of treads	=	3–8°
Slope of scarps	=	15–22°, with occasional vertical rocky outcrops

When traced laterally, few of the terraces maintain a horizontal attitude. Although
developed generally on the Dartmoor metamorphic aureole (and mainly in the height
range 300–450 m) the benches are in fact cut in a diversity of rock types, but all types
have one property in common, namely that they are well jointed. There are also some
definite structural relationships: for instance, the terraces are horizontal where they
lie along the strike. But their varied spacing, irregular distribution, the lateral slope
of some, and their association with evidence of periglacial action make it extremely
unlikely that they are the work of 'normal' subaerial, fluvial, or marine action: a periglacial interpretation raises fewer problems and allows them to be grouped with
morphologically similar features of present-day arctic regions, such as Waters describes
in Vest-Spitsbergen.

According to Waters, cryoplanation terraces originate as slight depressions in the
land surface, possibly joint-controlled, which become the sites for collection of moisture
and snow, and thus, under a cold climate, the sites of most intensive frost action. On
sloping ground, frost-broken material can be removed from the downslope edges of
any roughly transverse depressions by gelifluction and frost creep, once sufficient fine
material has accumulated by frost comminution. Gelifluction can carry away the
coarser fragments and allow frost-sapping to proceed further; the debris streams down
from one terrace to the next lower one only at selected points in the intervening scarp,
so that the latter is not buried by debris. However, on the lower parts of a hillslope,

the rate of accumulation of debris will be greater and bedrock will slowly be covered over completely. The upslope edge of initial transverse depressions is worn back by frost-sapping; at its foot, snow-banks will assist this process by supplying moisture, they will provide meltwater to wash away fines, and will encourage gelifluction across the tread below them. Other workers have added or emphasized other processes, such as cryoturbation on the flats, helping to sort and comminute the debris, and suffosion (piping) helping in the removal of fines by sub-surface wash. Cryoplanation terraces are thus envisaged primarily as bedrock-eroded features which at any given time will bear small and varying thicknesses of debris moving across them. Eakin was probably

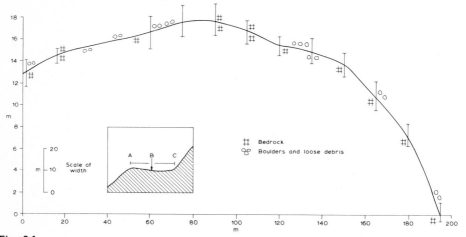

Fig. 6.1
Some features of a cryoplanation terrace by Butterfly Lake, west Baffin Island. The profile follows the median line of the terrace (point B on the inset.) The width of the terrace is indicated by short vertical lines, representing the distance A–C on the inset (A=outer edge of terrace, C=inner edge). Symbols above the profile refer to characteristics of the terrace surface in the inner B–C zone; below the profile they refer to the outer A–B zone. The altitude of the terrace varies lengthwise through 17 m and may be partly structurally controlled. The tread of the terrace slopes inward mostly, as indicated on the inset diagram

misled by this debris into regarding the terraces as constructional; the weight of evidence does not now favour this view.

Figure 6.1 shows some features of a cryoplanation terrace in western Baffin Island The terrace runs along a hillside by Butterfly Lake for about 200 m, its width varying between 7 and 15 m, and its height, measured along a median line, varying from 0 to 18 m above an arbitrary datum. Between points 1 and 12, the terrace surface has a reversed slope (that is, opposite to the general slope of the hillside), its inner edge being up to 1 m lower than its outer edge. The position of the terrace in this case appears to be structurally controlled. Nivation processes were thought to be the most important in the development of this terrace.

Demek (1969a) has provided a comprehensive survey of the world distribution, genesis and development of cryoplanation terraces. Some general climatic relationships emerge: permafrost is not essential to the formation of cryoplanation features, but there must be active freeze-thaw action and enough snow to supply meltwater in the summer. He claims (Demek, 1969b) that a sequence of developmental stages can be recognized:

1 Nivation hollows develop, as described in the previous chapter. If there are already irregularities in the ground surface, snow patches can at once accumulate. Otherwise, frost scars, polygonal cracks or the banks of gelifluction lobes can provide the loci for nivation. Strongly reflecting a nivational origin is the fact that most cryoplanation terraces are elongated and of sickle-shaped form, and are rarely found on very steep slopes where snow will not accumulate.
2 The initial hollow broadens to form a terrace or 'cryopediment' (the latter term is said to be appropriate since back-wearing of a steep slope at the rear is common to cryoplanation terraces and to pediments in general, and the surface of the terrace, again like that of any pediment, is essentially a surface of transport.
3 The cryopediments extend and coalesce or intersect, producing summit flats. On these, the action of running water, gelifluction and frost creep cease to be important because of the low gradient, and suffosion becomes dominant, carrying away the fines between and under the boulders. Possibly this is one way in which summit blockfields evolve, as H. T. U. Smith (1968) has suggested.

2 Tors

It should be clearly stated at the outset that there are many features which have been or can be described as tors whose formation has no connection with periglacial conditions or processes. Tor was first proposed as a scientific term by D. L. Linton (1955), and geomorphologically is used to designate a residual mass of upward projecting bedrock exposed by differential weathering and removal of weathered debris, though there is considerable variety in their form. They are most commonly associated with intrusive igneous, metamorphic and some sedimentary rocks such as massively jointed sandstone. Some mark resistant outcrops, others do not; they may occur in groups or singly: they are found on hill summits and on hill slopes, but rarely on valley floors. The joint-blocks of the tor may be loosely piled resembling 'the squared stones of cyclopean masonry' (R. H. Worth, 1930) or they may fit so closely that the blocks cannot have suffered any movement relative to one another. The blocks may be both rounded and angular. The tors frequently rise abruptly from a surrounding relatively smooth surface, which may be of bedrock (such as Linton's (1955) 'basal platforms') or may represent a surface of accumulation: Plate 8 in Eakin (1916) is a fine example of a granite tor surrounded by a smooth surface of gelifluction deposits.

It is not possible here to review all the various opinions expressed on the origin of tors, for the term has been used to describe a variety of features developing under conditions ranging from tropical to arctic. Attention will be confined to tors in areas which now experience, or have in the Pleistocene experienced, a severe frost climate. But even in these areas there is still controversy over whether the tors may be distinctive periglacial landforms, or whether they may be landforms relict from previous periods of totally different climate and later modified by periglacial conditions. Some workers admit both possibilities in the same general area. An example is Demek's work (1964)

Plate XVI
Tor exposed on the slopes of Jason's Creek, Devon Island, Canada. (C.E.)

on tors in the Bohemian upland of Czechoslovakia. Two possible models of their evolution are put forward. The first envisages two stages in their formation:

1 a period of deep weathering under warm humid conditions (possibly in the Tertiary) in which a thick regolith developed to 110 m in places. In this regolith, core-stones developed where joint planes were more widely spaced;
2 a period or periods of regolith removal under periglacial conditions in the Pleistocene, when the tors were first exhumed and subjected to frost attack and extensive mass wasting. This hypothesis is applied by Demek to many tors in western Bohemia.

Other tors are considered by Demek (see also T. Czudek and Demek, 1971) to have developed in one stage by rock weathering and simultaneous removal of debris in a periglacial environment. Some hillslopes consist of a series of cryopediments whose edges are marked by tors and frost-riven cliffs; at the foot of these, angular talus has accumulated in places and sometimes gives rise to stone streams. The cliffs retreat by frost attack, intersecting one another to produce tors and other related forms. This scheme of tor evolution is also put forward by Czudek (1964) in connection with the Hrubý Jeseník Mountains of northern Moravia (Fig. 6.2). The tors here, which attain heights of 20 m and sometimes possess overhanging faces, are thought to have formed

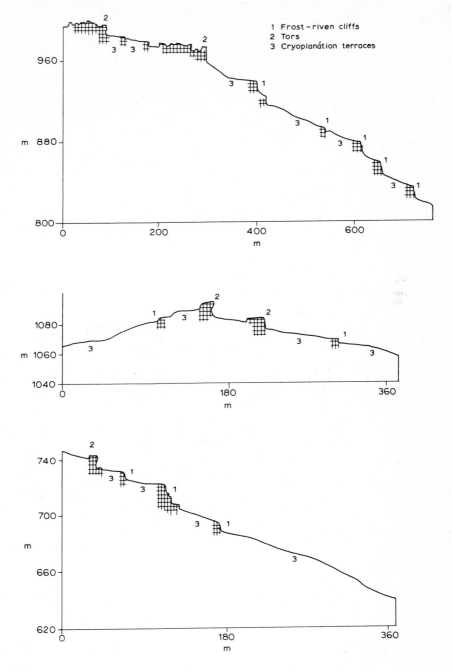

Fig. 6.2
Slope profiles in the Hrubý Jesenik Mountains, Czechoslovakia (T. Czudek, *Biul. Peryglac.*, 1964)

solely under periglacial conditions, there being no evidence of exhumation of Tertiary weathering forms.

A one-stage model is also advocated for tors in the McMurdo Sound area of Antarctica by M. J. Selby (1972) (Fig. 6.3). In this cold arid region, chemical weathering is nevertheless active (see Chapter 1), especially salt weathering, and the wind is an important agent of removal of the finest debris.

The origin of the tors in south-west England and in the gritstone country of the Pennines has been the subject of prolonged controversy. Linton (1955, 1964) has strongly maintained a non-periglacial origin, invoking deep weathering under warm climatic conditions and subsequent exhumation of tor forms, as in Demek's two-stage

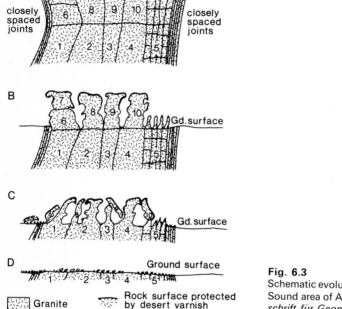

Fig. 6.3
Schematic evolution of tors in the McMurdo Sound area of Antarctica (M. J. Selby, *Zeitschrift für Geomorphologie*, 1972)

hypothesis. Periglacial mass movements have been mainly responsible for removing and redistributing loose debris already in existence and in displacing some of the core-stones. Gelifluction and frost action are held not to have assisted in producing tors but in destroying them. On the other hand, J. Palmer and J. Radley (1961), discussing Pennine tors, and J. Palmer and R. A. Neilson (1962), concerned with Dartmoor tors, argue that the tors were formed essentially under Pleistocene periglacial conditions. Palmer and Neilson maintain that on Dartmoor the tors are dominantly frost-broken and angular: any rounding of component blocks is said to be secondary. The *growan* or disintegrated granite, which they distinguish from the kaolinized granite, may be substantially the result of frost pulverization and gelifluction, and they contend that the *clitter* of granite blocks spread downslope from the tors is frost-broken material moved by gelifluction and frost creep. The processes were arrested by the ending of periglacial conditions, and thus the tors survive.

Waters (1965) attempts a compromise view in relation to south-west England. Selective bedrock decomposition in pre-glacial (possibly inter-glacial) warm periods is accepted; the effect of periglacial conditions has been to remove both finer debris and large blocks from the tors and their vicinity by gelifluction, and once exposed, the tors were subjected to destructive frost action, though it is very clear that some rocks were much more susceptible to frost action than others.

Derbyshire (1972) argues that the conjunction of rounded and angular tor forms in the same area is more easily and plausibly explained in terms of local differences in micro-climate, rather than by appealing to any two-stage theory of evolution. In southern Victoria Land, Antarctica, tors in sandstone and dolerite possess both rounded and angular features. Some of the rounding may be due to wind abrasion in this relatively arid region, but much of it is associated with contemporary chemical weathering, as shown by Derbyshire's studies of the weathering residues. Angular features probably produced by frost shattering predominate in moister sites, while chemical weathering appears to be more active on snow-free sites. Derbyshire's findings do not support Linton's hypothesis in this area, nor the hypotheses of Palmer, Neilson and Radley. The idea that tors of a certain shape are diagnostic of periglacial conditions seems to be untenable, nor is deep weathering a necessary pre-requisite for rounded tor formation.

Similar conclusions were reached by R. Dahl (1966) comparing tors in northern Norway and northern Italy. Dahl also noted how tors may be present in areas which until quite recently were covered by glacial ice; therefore tors cannot be used as evidence of unglaciated areas. This last point has been taken up by several geomorphologists and Dahl's views supported. D. E. Sugden (1968) working in the Cairngorm Mountains of Scotland finds that tors on plateau surfaces survived envelopment by slow-moving ice, occasionally with signs of slight ice moulding; similar conclusions by C. M. Clapperton (1970) in the Cheviot Hills of northern England reinforce the view that tors cannot be used as indicators of non-glaciated terrain.

In general, it is becoming clear that there is considerable variety of tor morphology in present and past periglacial regions, and that a single hypothesis cannot cover all cases. The clearest evidence on periglacial tor formation is provided by regions where these features are being actively produced at the present day, such as parts of northern Canada (Plate XVI) or of Antarctica. In these areas, there appear to be close links between tor morphology and the local micro-climatic conditions controlling the weathering processes.

3 Blockfields and blockstreams

Accumulations of coarse detritus on level or gently sloping mountain-top surfaces have been termed blockfields, Felsenmeere or Blockmeere. Washburn (1973) suggests that the adjective 'block' might be reserved for predominantly angular accumulations, though the degree of angularity may be as much a function of rock type as of process. Sometimes, indeed, the boulders in blockfields can be misleadingly round, giving the appearance of a very stony moraine. The processes producing blockfields are mainly connected with a freeze-thaw environment, now or in the past: G. M. Richmond (1962) in the La Sal Mountains of Utah cites frost creep, frost wedging and frost sorting. Eluviation of fines is another factor (see the reference to piping and suffosion on p. 163). But it is also important to note that non-periglacial weathering processes can

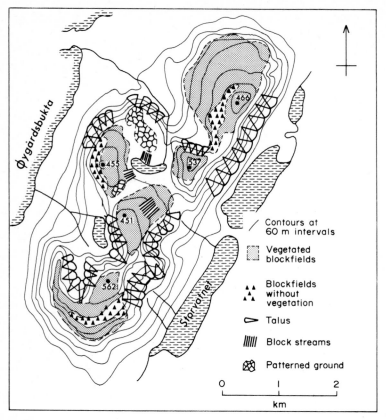

Fig. 6.4
Geomorphological map of a small area (6×3 km) in northern Scandinavia showing blockfields, block-streams, talus and patterned ground (L. Strömqvist, 1973, *Uppsala Universitet Naturgeografiska Institutionen*, Report 22)

produce angular material: mechanical failure of a well-jointed stratum overlying incompetent beds or as a result of piping and the development of dilatation joints are possibilities. On the other hand, large accumulations of angular detritus in areas where such debris is not forming today (as shown, for instance, by lichens and vegetation growing on it) are certainly strong pointers to former frost action.

Fig. 6.4 shows an area in northern Scandinavia mapped by L. Strömqvist (1973), including blockfields with and without a vegetative cover. The block mantle in this and two other sample areas was found to be remarkably uniform in thickness at about 0·7 m. This was apparently independent of rock type, local climate and age, but with increasing gradient and transition from blockfields to blockstreams (see later), the mantle thinned and larger blocks became more noticeable. Strömqvist considers that the accumulations originated from well-jointed bedrock. Frost action was the primary process, favoured by temperature régime and abundance of moisture.

Blockfields may consist entirely of local material, or it is possible that rock debris from more distant sources has been incorporated by glacial ice. Fabric as well as lithological studies provide a possible means of distinction between undisturbed autochthonous blockfields and those that have modified by overrunning ice. In the former, the preferred orientation should theoretically run parallel to the local slope, as with other slope deposits, but on flat or very gently sloping surfaces, the fabric may be weak or irregular. N. Caine (1968a, b, 1972) has examined the fabrics of blockfield material in Tasmania (Fig. 6.5). Air photo analysis showed that, in general, there was a strong tendency for the *a* axis to be aligned within 10° of the slope direction, and the vector strength often exceeded the 95 per cent significance level on the Rayleigh test; the inclination of the long axes varied between horizontal and the local slope angle usually. For some blockfields, the pole of the fabric was skewed slightly relative to the slope direction, possibly owing to impedance of movement which in turn might be related to the removal of interstitial material by piping. At the edges of blockfields or the toes of blockstreams, the fabric may become transverse. Caine also noted that very large blocks tended to distort the flow pattern.

The distribution of blockfields, like tors, has been used to indicate limits of ice cover, but caution should be exercised in this respect (see J. D. Ives, 1958, for instance) for it appears possible for such detritus to remain relatively unaffected by the passage of ice over it. The length of time required for blockfields to form varies considerably with rock type, and also with climate, especially the number of significant oscillations through freezing-point (frost cycles). R. Dahl (1966) finds evidence that some blockfields in Nordland, Norway, have developed in the post-glacial, and therefore they cannot be used to indicate possible ice-free areas in the last glaciation.

C. A. M. King and R. A. Hirst (1964) contrast blockfields with boulder fields of marine origin in the Åland Islands off south-west Finland. The different types, the one of angular blocks and the other of rounded boulders, occur in close proximity, and care must be taken to distinguish them. The angular blockfield (Fig. 6.6A) is 110 m long and varies in width from 55 to 21 m, with a mean gradient of 1 in 29. The margins of the blockfield provide useful evidence on the method of its formation. The loose blocks, near the upper edge of the blockfield, can be traced into a zone of greatly weathered and cracked slabs *in situ*. These slabs in turn merge into solid bedrock; this consists of a type of Rapakivi granite, which breaks up very readily along vertical joints and horizontal planes of weakness. At the lower end of the blockfield, the angular blocks of local granite overlie smaller and rounder stones, which include some erratics. This suggests that the angular stones have moved downslope to cover the older deposits. It is thought that this blockfield originated largely by frost shattering of the well-jointed granite. Frost heaving has subsequently assisted in the downhill movement of the blocks, and the fact that the blockfield faces south would facilitate alternate thawing and freezing during spring and autumn. It is considered likely that this blockfield has been formed by the action of frost during the relatively short period since 4000 BC

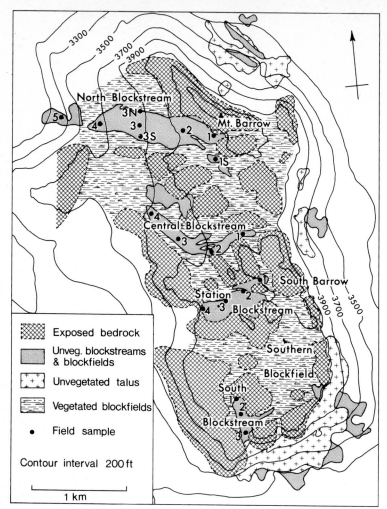

Fig. 6.5
Map of blockfields and blockstreams on Mount Barrow, Tasmania (N. Caine, *Geografiska Annaler*, Stockholm, 1968).

when this part of the island emerged from the sea.

The rounded boulder field (Fig. 6.6B) shows several significant differences from the angular one. It consists of well-rounded stones that are rather larger on conspicuous ridges that extend across the field parallel to the contours. The slope of the boulder field, about 1 in 10, is much steeper than that of the blockfield. The rounded boulder field consists of two parts, one facing north-east and the other south-west, both with well-marked ridges. The margins of the boulder field are again significant. Around its upper parts, the boulders rest discordantly against smoothed and solid outcrops of granite. A sharp boundary delimits the upper part of the boulder field, but its lower margin merges into the forested ground as the bare rocks become gradually better

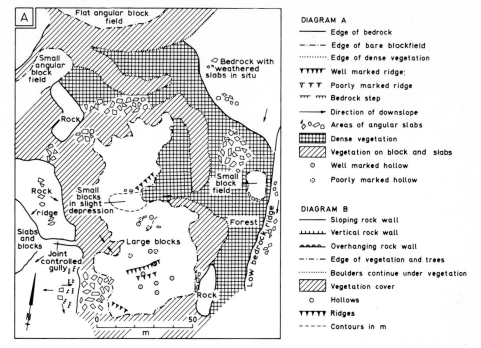

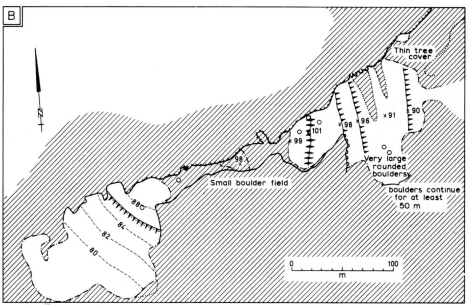

DIAGRAM A

——————— Edge of bedrock
—·—·— Edge of bare blockfield
············· Edge of dense vegetation
ᴛᴛᴛᴛᴛ Well marked ridge
ᴛ ᴛ ᴛ Poorly marked ridge
ᴛᴛᴛ ᴛᴛᴛ Bedrock step
———→ Direction of downslope
◌◦◯◦◯◦ Areas of angular slabs
▦ Dense vegetation
▨ Vegetation on block and slabs
○ Well marked hollow
◌ Poorly marked hollow

DIAGRAM B

——————— Sloping rock wall
⊥⊥⊥⊥⊥⊥ Vertical rock wall
∿∿∿∿ Overhanging rock wall
—·—·— Edge of vegetation and trees
············· Boulders continue under vegetation
▨ Vegetation cover
○ Hollows
ᴛᴛᴛᴛᴛ Ridges
— — — — Contours in m

Fig. 6.6
A Map of an angular blockfield, Åland Islands, Finland;
B Map of a rounded boulder field, Åland Islands, Finland (C. A. M. King and R. A. Hirst, *Fennia*, by permission of the Johnson Reprint Corporation)

covered with soil and vegetation. At the upper limit of the north-east facing boulder field, there is a marked summit ridge of large boulders, with a backslope leading down through a gully to the lower south-west facing boulder field. Unlike the blockfield, there are numerous well rounded erratics on its surface which must have been carried into the vicinity by ice. The character of the margins, the size distribution of the stones and the gradient of the boulder field all support the suggestion that these rounded boulder fields were formed not under periglacial conditions but by storm wave action.

Table 6.1 Relationship between block accumulations and gradient

Gradient	Block orientation	Micro-relief	Dominant processes
0–10°	Increasing number orientated downslope	Polygons (0°) merging into stripes (10°)	Cryoturbation; some gelifluction
10–20°	Strong orientation downslope	Sorted steps	Gelifluction and frost creep
20–35°	Fewer blocks aligned downslope; imbrication developing	Patterned ground diminishing in importance	Increasing slope wash. Blocks accumulating against each other
Over 35°	Blocks mainly orientated perpendicular to the slope	No patterned ground	Block creep. Most fine-grained interstitial material removed

This must have taken place soon after the island emerged from the sea about 6000 BC, owing to isostatic recovery, and the fact that the boulder field lies at about 100 m in altitude is in agreement with this date. At this time, the fetch would have been considerable in a direction perpendicular to the boulder ridges. The coarse nature of the boulder fields would allow rapid percolation of water and hence the minimum amount of later frost disturbance of these wave-formed features.

3.1 *Blockstreams*

Blockstreams (or block slopes) are distinguished from blockfields solely in terms of gradient. The distinction is therefore arbitrary. Washburn (1973) suggests an angle of 5° as a suitable separating gradient, but J. Piirola (1969) uses the term blockfield on much steeper slopes: the mean gradient of the largest blockfields in an area in western Inari, Finland, is over 11°, while blockstreams attain angles of 25°. On still steeper slopes, the blocks roll and slip more freely and talus formations develop with pronounced size sorting. Table 6.1 shows a classification condensed from Strömqvist (1973) and based on gradient.

The margins of blockstreams are often ill-defined, for the surrounding slopes may also possess many boulders, and the distinction then may lie solely in the presence of more soil and vegetation on the slopes and the relative lack of soil and vegetation on the blockstreams themselves.

Fossil (Pleistocene) blockstreams are known in a few parts of southern Britain beyond the limits of the last glaciation. R. B. G. Williams (1968) notes the following examples, together with their lengths and average gradients:

Stow-cum-Quy, Cambs.	10·3 km	0° 30′
Below Hythe Beds scarp in W. Kent	6·4 km	< 1°
Clatford Bottom, Wilts. (sarsens)	4·0 km	1° 30′
Lockeridge Dene, Wilts. (sarsens)	3·2 km	1° 30′
Valley of Stones, Dorset	1·8 km	2°
Leusden Common, Dartmoor	0·9 km	7° 30′
Yarner Woods, Dartmoor	0·8 km	11° 30′

3.2   *The stone runs of the Falkland Islands*

Ever since Sir Charles Darwin described the 'stone river' at the head of Berkeley Sound in the Falkland Islands in 1846, attributing it to some catastrophic earthquake, these features variously termed stone runs or block cascades have attracted attention and controversy over their origin. C. W. Thomson (1877) was the first to suggest slow downhill creep of the blocks as water washed out fine debris, and by alternate soaking and drying of the ground beneath the blocks. J. Geikie (1894) compared the stone streams of the Falklands to the 'rubble drift' of southern England formed under conditions of alternate freezing and thawing.

All workers now regard them as of periglacial origin and in the same general category as blockstreams, though the mechanism of their formation and possible movement is by no means clear. One hypothesis that commanded much support was that of J. G. Andersson (1906), who suggested that the blocks in the stone streams of the Falkland Islands had been borne along down-valley by flowing mud and fine debris at a time of severely cold climate. The Falkland Islands were never ice-covered, but lay at times in the Pleistocene near the margin of the Antarctic ice. Frost action produced the debris, gelifluction moved it, and later, running water washed out the fine materials from the upper layers but was itself unable to move the blocks. J. Büdel (1937) adopted a similar view for blockstreams in the German Mittelgebirge, and also H. T. U. Smith (1949, 1953) in Wisconsin.

R. P. Sharp (1942) in the Yukon was able to observe mudflows in the thaw season carrying boulders up to 1·5 m in diameter, for mud has a much higher specific gravity than water. On the other hand, Washburn (1947) thought the blockstreams he examined in northern Canada were not related to the viscous flow of an underlying fine-grained layer but to any factors (such as frost-heave or slope-wash) which would upset the equilibrium of blocks resting on a slope.

J. R. F. Joyce (1950) has recently re-examined the stone runs of the Falkland Islands, and finds Andersson's solifluction hypothesis untenable. The component blocks are sometimes of enormous size and would have jammed immovably on gentle slopes—some stone rivers here stretch for 4 km on grades of a few per cent only (compare the figures given by Williams above). The underlying fine-grained layer demanded by Andersson is not to be found, nor is it at all evident where it could have come from in sufficient volume to convey the vast bulk of huge stones. Furthermore, some of the stone runs lie not in valleys but are really blockfields across low rounded plateaux. A significant fact is that the stone runs consist of quartzite boulders derived solely from one highly jointed unit of the Devono-Carboniferous series. Joyce argues that the boulders have not in fact travelled any distance but were formed largely *in situ* in the Pleistocene on an outcrop underlain by shale and very susceptible to frost weathering. As the bounding scarps of the quartzite beds retreated under frost attack, so a litter of angular blocks accumulated to carpet the area over which the scarps retreated. Vegetation has since been unable to gain a foothold because of the size of the interstices and thus they remain as barren stone runs. In this case, blockfield is a more appropriate term than stone stream, for the latter conveys too strong an impression of extensive downslope movement, which Joyce denies.

The matter has been recently re-examined by R. Clark (1972), who notes that Joyce tended to confuse stone runs with hill-top blockfields. Clark recognizes an interrelated progression of features from hill tops to valley floors. The high ground is characterized by blockfields and cryoplanation surfaces, together with some residual tors. On slopes, frost-riven cliffs, talus, blockstreams and terraces dominate, while stone runs and head deposits (into which the stone runs merge as they pass into coastal areas of low gradient) occupy the valley floors.

It is thus evident that there are many types of coarse detrital accumulations, from summit blockfields to stone runs and blockstreams. Only further field studies can serve to differentiate them more satisfactorily in individual areas, and the terminology has suffered from great confusion in the past (for instance, Kesseli's use of the term 'rock stream' to describe rock glaciers in the Sierra Nevada: see p. 122. Frost action is the fundamental process of derivation of the blocky debris: frost-heave, gelifluction, and any other processes that assist gravity by disturbing the equilibrium of large blocks on sloping surfaces, have been responsible for whatever movement of the boulders has taken place. It seems ·clear that the earlier workers had exaggerated notions of the extent of movement on the more gentle slopes, and gelifluction should not be too readily invoked in the absence of any fine-grained material lying beneath the blocks. Another method by which some blockfields may form is by the decay of rock glaciers. Melting of interstitial ice will leave behind spreads of rocky detritus, incapable of further motion.

4 Conclusion

Several features related to the work of frost action, nivation and gelifluction character-ize areas of past and present periglacial activity. Cryoplanation terraces represent bedrock-eroded features, often structurally controlled, in whose formation frost action and nivation play a major part. They occur at irregular vertical intervals and com-monly vary in altitude as they are traced along a hillslope. Snow-melt and gelifluction are important in the removal of debris from them. Their edges are marked by frost-riven cliffs and tors in some cases.

The origin of tors is still highly controversial. It is clear, however, that by no means all tors are of periglacial origin, and some workers deny any strong connection between tors and periglacial processes. On the other hand, there is evidence that tors in parts of Britain and Czechoslovakia, for instance, have been extensively modified, and possibly even formed, under Pleistocene periglacial conditions. The degree of rounding or angularity of the tor features seems in most cases to reflect differences of weathering process related to micro-climate, and it is not necessary to invoke two-stage theories of origin. Neither are tors any reliable guide to the extent of former ice cover.

Accumulations of coarse debris in the form of blockfields, blockstreams and stone runs are in the majority of cases related to cold climatic conditions, though rounded boulder fields of marine origin must be carefully differentiated. The angular blocks of the periglacial accumulations are initially derived by frost action, and on sufficiently sloping surfaces may thereafter be slowly displaced by frost-heave, gelifluction, or the washing-out of any fine interstitial or subjacent debris which would disturb the equilibrium of the blocks. Large-scale gelifluction as envisaged by Andersson, however, is unlikely.

5 References

ANDERSSON, J. G. (1906), 'Solifluction, a component of sub-aerial denudation', *J. Geol.* **14**, 91–112

BRYAN, K. (1946), 'Cryopedology—the study of frozen ground and intensive frost action, with suggestions on nomenclature', *Am. J. Sci.* **244**, 622–42

BÜDEL, J. (1937), 'Eiszeitliche und rezente Verwitterung und Abtragung im ehemals nicht vereisten Teil Mitteleuropas', *Petermanns Mitt., Ergänz.* **229**, 71 pp.

CAINE, N. (1968a), 'The blockfields of north-eastern Tasmania', *Aust. Nat. Univ., Dept. Geogr. Publ.* **G6**, 127 pp.

(1968b), 'The fabric of periglacial blockfield material on Mt. Barrow, Tasmania', *Geogr. Annlr* **50**A, 193–206

(1972), 'Air photo analysis of blockfield fabrics in Talus Valley, Tasmania', *J. sedim. Petrol.* **42**, 33–48

CAIRNES, D. D. (1912), 'Differential erosion and equiplanation in portions of Yukon and Alaska', *Bull. geol. Soc. Am.* **23**, 333–48

CLAPPERTON, C. M. (1970), 'The evidence for a Cheviot ice cap', *Trans. Inst. Br. Geogr.* **50**, 115–27

CLARK, R. (1972), 'Periglacial landforms and landscapes in the Falkland Islands', *Biul. Peryglac.* **21**, 33–50

CZUDEK, T. (1964), 'Periglacial slope development in the area of the Bohemian massif in northern Moravia', *Biul. Peryglac.* **14**, 169–93

CZUDEK, T. and DEMEK, J. (1971), 'Pleistocene cryoplanation in the Česka vysocina highlands, Czechoslovakia', *Trans. Inst. Br. Geogr.* **52**, 95–112

DAHL, R. (1966), 'Block fields, weathering pits and tor-like forms in the Narvik Mountains, Nordland, Norway', *Geogr. Annlr* **48**, 55–85

DARWIN, C. (1846), *Geological observations on South America* (London), **7**, 279

DEMEK, J. (1964), 'Castle koppies and tors in the Bohemian highland (Czechoslovakia)', *Biul. Peryglac.* **14**, 195–216

(1969a), 'Cryoplanation terraces, their geographical distribution, genesis and development', *Rozpr. čsl. Akad. Ved., r. M.P.V.* **79**, 1–80

(1969b), 'Cryogene processes and the development of cryoplanation terraces', *Biul. Peryglac.* **18**, 115–25

DERBYSHIRE, E. (1972), 'Tors, rock weathering and climate in southern Victoria Land, Antarctica', *Inst. Br. Geogr. Spec. Publ.* **5**, 93–105

EAKIN, H. M. (1916), 'The Yukon-Koyukuk region, Alaska', *U.S. geol. Surv. Bull.* **631**, 1–88

GEIKIE, J. (1894), *The Great Ice Age* (London, 3rd Ed.)

GUILCHER, A. (1950), 'Nivation, cryoplanation et solifluction quaternaires dans les collines de Bretagne occidentale et du Nord de Devonshire', *Revue Géomorph. dyn.* **1**, 53–78

IVES, J. D. (1958), 'Mountain-top detritus and the extent of the last glaciation in north-east Labrador-Ungava', *Can. Geogr.* **12**, 25–31

JORRÉ, G. (1933), 'Le problème des "terrasses goletz" sibériennes', *Revue Géogr. alp.* **21**, 347–71

JOYCE, J. R. F. (1950), 'Stone runs of the Falkland Islands', *Geol. Mag.* **87**, 105–15

KING, C. A. M. and HIRST, R. A. (1964), 'The boulder fields of the Åland Islands', *Fennia* **89**, 1–41

KOZMIN, N. M. (1890), quoted by DEMEK, J. (1969a)

LINTON, D. L. (1955), 'The problem of tors', *Geogrl J.* **121**, 478–87
(1964), 'The origin of the Pennine tors—an essay in analysis', *Z. Geomorph.* NF **8**, 5–24

PALMER, J. and RADLEY, J. (1961), 'Gritstone tors of the English Pennines', *Z. Geomorph.* NF **5**, 37–52

PALMER, J. and NEILSON, R. A. (1962), 'The origin of granite tors on Dartmoor', *Proc. Yorks. geol. Soc.* **33**, 315–40

PIIROLA, J. (1969), 'Frost-sorted block concentrations in western Inari, Finnish Lapland', *Fennia* **99**(2), 35 pp.

PRINDLE, L. M. (1905), 'The gold placers of the Fortymile Birch Creek and Fairbanks regions', *U.S. geol. Surv. Bull.* **251**, 89 pp.

RICHMOND, G. M. (1962), 'Quaternary stratigraphy of the La Sal Mountains, Utah', *U.S. geol. Surv. Prof. Pap.* **324**, 135 pp.

SELBY, M. J. (1972), 'Antarctic tors', *Z. Geomorph.*, Suppl. **13**, 73–86

SHARP, R. P. (1942), 'Mudflow levees', *J. Geomorph.* **5**, 222–7

SMITH, H. T. U. (1949), 'Periglacial features in the driftless area of southern Wisconsin', *J. Geol.* **57**, 196–215
(1953), 'The Hickory Run boulder field, Carbon County, Pennsylvania', *Am. J. Sci.* **251**, 625–42
(1968), 'Piping in relation to periglacial boulder concentrations', *Biul. Peryglac.* **17**, 195–204

STRÖMQVIST, L. (1973), 'Geomorfologiska studier av Blockhav och Blockfält i norra Skandinavien', *Uppsala Univ. Naturgeogr. Inst. Rep.* **22**, 161 pp.

SUGDEN, D. E. (1968), 'The selectivity of glacial erosion in the Cairngorm Mountains, Scotland', *Trans. Inst. Br. Geogr.* **45**, 79–92

TE PUNGA, M. T. (1956), 'Altiplanation terraces in southern England', *Biul. Peryglac.* **4**, 331–8

THOMSON, C. W. (1877), *The Atlantic* (London) **2**, xi, 396

WASHBURN, A. L. (1947), 'Reconnaissance geology of portions of Victoria Island and adjacent regions, Arctic Canada', *Am. geol. Soc. Mem.* **22**, 142 pp.

(1973), *Periglacial processes and environments*

WATERS, R. S. (1962), 'Altiplanation terraces and slope development in Vest-Spitsbergen and south-west England', *Biul. Peryglac.* **11**, 89–101

(1965), 'The geomorphological significance of Pleistocene frost action in south-west England' in *Essays in Geography for Austin Miller* (ed. J. B. WHITTOW and P. D. WOOD, 39–57

WILLIAMS, R. G. B. (1968), 'Some estimates of periglacial erosion in southern and eastern England', *Biul. Peryglac.* **17**, 311–35

WORTH, R. H. (1930), 'The physical geography of Dartmoor', *Rep. Trans. Devon. Ass. Advmt Sci.* **62**, 49–115

7

Periglacial wind action

The winds blew ... during a Dust Age over outwash fans and plains ... exposed and dried during the intervals between successive inundations. (J. K. CHARLESWORTH, 1957)

Wind action under cold dry conditions is responsible for the most widespread and bulky of all periglacial deposits, namely loess, though its eolian origin was not universally accepted in the past. The loess associated with regions which lay just outside the embrace of the last glaciation suggested further searches for other forms resulting from periglacial wind action, for loess represents but the finer fraction of wind-transported debris, and the strength of wind action that it demanded ought to be independently supported by wind erosion forms. Deposits of coarser disseminated wind-blown sand were identified in the 1940s, fossilized dune forms have been described and analysed since the early part of this century, while the effects of sand-blasting are now well documented in the form of ventifacts and wind-worn rock surfaces. All these features are not only of interest in themselves, reflecting a particular aspect of periglacial activity, but yield valuable evidence on the climatic conditions prevailing at certain times in the Pleistocene.

1 Wind erosion

Wind erosion is manifested in the occurrence of faceted, fluted and grooved surfaces on bedrock, but more commonly on pebbles and boulders projecting up from the ground surface. Ventifacts related to strong wind action of Pleistocene time are found today lying exposed on surfaces now no longer subject to such attack, or else buried by contemporaneous or later eolian deposits. Examples of buried ventifacts, sometimes in numbers sufficient to justify their description as 'desert pavement', are represented by the Steinsohle beneath the loess in north Germany, or by the wind-blasted stones beneath loess described by G. Johnsson (1958) in the Lund region of southern Sweden. In the Netherlands, W. Schönhage (1969) notes how ventifacts occur almost solely in desert pavement layers. These layers are typical deflation horizons, composed of a gravelly lag deposit within a fine sandy matrix, which truncate the underlying deposits along a relatively horizontal surface. They may occur at several levels in the sequence of late-Pleistocene wind-blown deposits. R. Paepe and A. Pissart (1969) describe how some in Belgium may be traced over 5 km or more; equally extensive occurrences are noted in central Poland by J. Dylik (1969). In the United States, C. K. Wentworth and R. I. Dickey (1935) summarized the then known ventifact localities, and W. E. Powers (1936) provided some additional occurrences. Powers, for instance, recorded beautifully and unmistakably wind-grooved boulders lying on the sandy plain of former Lake Wisconsin; and similar ventifacts and grooved boulders of South Park, Colorado, enabled him to establish that the dominant direction of the winds here responsible was from the north-west, mainly in pre-Wisconsin glacial periods.

Ventifact formation generally requires strong winds, open topography that does not hinder their action, minimal vegetative ground cover, and an adequate supply of suitable sand grains for cutting. Ph. H. Kuenen's (1960) wind-tunnel experiments showed that ventifacts form only very slowly with moderate wind speeds and fine sand grains, but, at the other extreme, with coarse angular grains blowing at storm velocities, periods of a few weeks can suffice. The lower size limit for the grains responsible for the sand-blasting is 0·05 mm (in the case of quartz grains), partly because grains finer than this travel in suspension, not by saltation, and therefore rarely impinge on pebbles or rock at the ground surface. R. P. Sharp (1949) discusses some of the problems involved in determining former wind directions from ventifacts, with reference to the Big Horn region of Wyoming where large areas of ground are littered with ventifacts ranging from less than 2–3 cm in size to boulders 2 m in length. The smaller ventifacts may display up to twenty wind-cut faces and have clearly moved during sand-blasting; but the boulders more than 0·3 m in size and with but a single wind-cut facet are thought to be *in situ* and provide reliable indications of wind-direction. If wind-cut grooves on the boulders are to be used for this purpose, it must be remembered that the wind will itself be deflected by a large boulder; therefore, only grooves following the true dip of a wind-cut facet should be so used. The most probable causes of the shifting of the smaller stones during wind attack are either the wind itself or frost action (A. Cailleux, 1942). Sharp regards frost-heave or gelifluction as most likely and deduces from this that the cutting of the ventifacts occurred in cold periods. Unmoved boulders from some thirty localities in the Big Horn region show winds blowing from directions between N.10°W and N.60°W, essentially the same as today and reflecting the dominant effect of the mountain barriers.

R. L. Nichols (1969) records numerous well-developed ventifacts of the present-day in north-west Greenland. On the windward faces of boulders, sand-blasting, which occurs mainly in summer when the ground is not snow-covered, is efficient at cleanly removing limonite or other weathering stains. The strong winds in this area have also truncated and grooved earth and turf hummocks.

A final comment relevant to periglacial wind erosion is that blowing snow can have a similar geomorphological effect at very low temperatures. Below −40°C, snow crystals have a hardness similar to that of sand grains, and in the absence of vegetation, corrasion forms may develop in soft rocks. In high polar regions, particularly Antarctica, blowing snow has probably been underrated as an agent of erosion.

2 Disseminated eolian deposits

2.1   *Sand*

In discussing Pleistocene eolian deposits, it is useful to draw a rough distinction between silt (grain-size 0·004–0·06 mm) and sand (0·06–1·00 mm). Most work on sand deposits has in fact concerned the grain-size range 0·25–1·00 mm (medium-coarse); and R. A. Bagnold (1941) has shown that sand movement by saltation is most effective within this range. Wind-blown sand is found in many glacial deposits and sometimes forms a thin separate mantle over the ground surface when it is termed 'coversand'. J. Pelisek (1963) notes how such a mantle, with grain sizes dominantly in the range 0·25–0·50 mm, is common in the Bohemian lowlands and Carpathians, sometimes several metres thick. In the Netherlands, apart from the reclaimed polders and some small areas of boulder clay and loess in the north and east, a cover of yellow eolian sand of Weichselian age is widespread. The coversand, up to a few metres thick, may include niveo-eolian material (see p. 190), comprising sand and snow grains blown along together, the sand grains collapsing as the entrapped snow melts. Grain sizes are usually in the range 0·1–0·4 mm. The surface of the coversand is almost flat, or else possesses low arcuate ridges 1–2 m high and up to 100 m wide. East of Amersfoort there are straight narrow ridges up to 2·5 m high and 5 km long. In other places, the coversand has been reworked to form dune fields (G. C. Maarleveld, 1960). Horizontal bedding is visible in places, though stratification is generally poorly developed. Radiocarbon dating and pollen analysis of buried soils and peat within the coversands has related these organic layers to milder Weichselian intervals. The coversands extend into Belgium and north Germany, and it is possible to trace their transition through sandy loess to the true fine-grained loess.

Wind speeds involved in transporting such sand deposits must have been 30 km per hour or more. Sand grains will be rounded if they have been subjected to a long period of saltation, otherwise they will be sub-angular. A. Cailleux's (1942) study of the Pleistocene coarse sand deposits of Europe is still the most comprehensive yet undertaken. From about 3000 samples of various Quaternary deposits, the sand in the range 0·4–1·0 mm was separated, and the percentage of grains possessing a rounded and 'frosted' surface was recorded. Cailleux argued that matt or 'frosted' surfaces on the sand grains were indicative of eolian abrasion, but this is not now thought to be a completely reliable indicator. Kuenen and W. G. Perdok (1962) have shown experimentally that chemical frosting may occur due to corrosive solutions and also by mechanical abrasion during water transport. M. Seppälä (1969) ascribes the matt or semi-matt surfaces of wind-blown sand grains in Finland to frost weathering.

Fig. 7.1 shows the general distribution of frosted, supposedly wind abraded, sand grains in Europe. It does not refer to any particular interval of the Pleistocene. Cailleux (1969) has subsequently extended his survey into the Soviet Union, tracing the belt of strong periglacial wind action from central Poland towards the Moscow region and northwards to the Timan Mountains. Other separate areas identified in the Soviet Union are the Ob and Lena basins. In central Poland, the percentage of wind-worn grains rises towards 100, and there is clearly revealed a belt of country lying just outside the Pomeranian ice limit in which there occurs a very high frequency of wind-worn grains. Within the belt, ventifacts are relatively abundant, and beyond it lies the main zone of deposition of finer wind-blown silt or loess. Cailleux argues for strongest wind action where the proportion of frosted grains is highest, and simultaneously, a severe periglacial climate, since the larger ventifacts in this area show signs of having been overturned by frost-heaving. Such conditions probably prevailed in the areas marginal

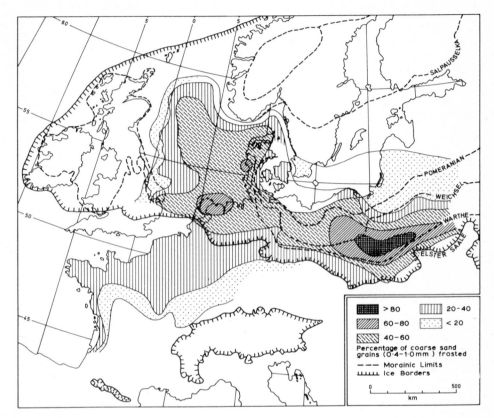

Fig. 7.1
Distribution of frosted (wind-worn) sand grains in the Quaternary deposits of Europe (A. Cailleux, *Memoir* 46, 1942, Geological Society of France)

to the Scandinavian ice in each of the older glacial periods, but the evidence of wind action is clearest (because best preserved) in the area bordering the last ice to reach the southern edge of the Baltic.

In Finland, Seppälä (1969) finds that the wind-blown sand grains tend to be more angular than those of central Europe. He suggests that the Finnish samples have been derived more recently by frost and glacial abrasion from the bedrock and that they have only been transported relatively short distances by wind. This may be true of Scandinavia generally.

2.2   *Silt*

Whereas sand is moved by wind primarily by the process of saltation near the ground, silt-sized particles may be lifted up to great heights (3 km or more) by eddy currents and will be transported for much greater distances before settling. Once deposited, however, wind-blown silt is unlikely to be again removed by wind owing to the cohesive effect between particles smaller than about 0·06 mm, unless the deposit is bombarded by saltating sand grains. The term 'loess' has been employed for eolian silt deposits since the earlier part of the last century, although it is now known that not all deposits formerly classified as loess are of purely eolian origin. It is unfortunate that, although the term loess is widely used, agreement is far from being reached on its definition. The International Quaternary Association's *Commission on Loess* is considering that it be defined as 'material with grain size between 2 and 60 μm', but this does not overcome the problem of differentiating between eolian, fluvial or other possible origins for loess and loess-like sediments. Moreover, by no means all eolian loess is related to periglacial conditions, for much loess has formed in relatively warm climates around the margins of some present-day deserts. Attempts have been made to distinguish between periglacial and desert loess, but the distinction is not a simple one for the loess of some regions in central Asia, for example, can be said to fall into both categories. The thickest deposits of loess in the world in north China (where a thickness of 600 m has been claimed; V. V. Popov, 1959) have often been regarded as desert loess, but the great bulk of them are of Pleistocene age and differ mainly in the magnitude of their development from the loesses of central Europe, for instance, most of which definitely accumulated under cold periglacial conditions. Periglacial and desert loess should not therefore be differentiated on a regional basis; much more significant is the fact that periglacial loess is better sorted, possessing a narrower range of grain size, for it was mostly derived from deposits already sorted and laid down by glacial meltwater. C. Troll (1944) argues that periglacial loess is also derived from frost-pulverized debris, for the ultimate effect of comminution by frost action is to produce material of grain size 0·01–0·2 mm, and this grain size range is also the commonest range for periglacial eolian sand and loess. He adopts the extreme view that all loess demands a cold climate—without frost, no loess. Kuenen (1960) has shown experimentally that the lower limit of size to which wind-blown quartz grains can be reduced by eolian action is 0·05 mm, yet the bulk of loess is finer than this. This confirms that the fine grains were already available to the wind, either as Troll suggested by frost pulverization, or from glacial rock flour, fine outwash or lacustrine deposits.

(*a*) *Composition of loess*   Fig. 7.2 shows typical graphs of grain-size distribution for samples of Kansan loess (A. Swineford and J. C. Frye, 1945). Silt-sized particles are dominant—normally, between 70 and 85 per cent by weight lie in the range 0·07–0·003 mm—but there is also a clay fraction with particles smaller than 0·002 mm which has only become amenable to analysis with the introduction of X-ray diffraction techniques (see, for example, J. C. Frye *et al.*, 1962). Some fine or very fine sand is also usually present. The grains are angular or sub-angular, but occasionally rounded. The chemical composition of loess is somewhat variable, especially in regard to its carbonate content which may range from nil to over 40 per cent by weight. A low $CaCO_3$ content reflects subsequent leaching. The dominant mineral in loess is quartz (50–80 per cent by weight); felspar may account for 20–25 per cent. The grains are bound together by clay minerals of which montmorillonite is often the most important—it constitutes up to 70 per cent of all clay minerals in Mississippi valley loess. Heavy mineral suites are often distinctive and may afford a means of identifying and correlating particular loesses: the same applies to the clay mineral suites (J. C. Frye *et al.*, 1962).

The buff colour of loess exposures is well known, though in the unweathered state (which is rare), the colour is grey. True loess is unstratified, for deposition was slow and each addition was an extremely thin layer. It is characterized by a vertical structure composed of numerous tubes representing the casts of former roots, for as the loess thickened, so the deeper parts of plant roots slowly decayed. The fine grain size and

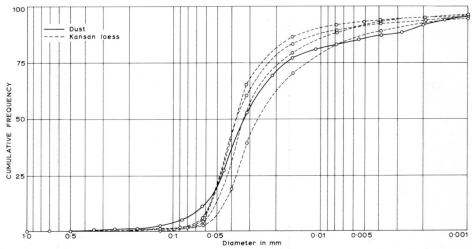

Fig. 7.2
Grain-size distribution of samples of Kansan loess and recent wind-blown dust (A. Swineford and J. C. Frye, *Am. J. Sci.*, 1945).

tubular structure result in a very high porosity—40 per cent is common and 65 per cent not unusual—so that the surface of loess remains dry and is relatively immune to dissection by water erosion. Wherever it is cut into by through-flowing streams, however, the loess is capable, because of its cohesive nature and vertical structure, of maintaining precipitous faces, such as appear in the well-known loess bluffs of the Mississippi. F. von Richthofen's (1877) classic description of the loess of north China is worth repeating:

> Loess is brown-yellowish and so crispy that it is easy to tear apart with the fingers, though solid enough—even where it has been attacked by destroying forces as, for example, running water—to stand with steep walls several hundred feet high. It is so fine-grained that it can be rubbed into the pores of the skin, whereafter only a few sand grains are left, and these are angular and without traces of rounding. By repeated washing, this sand can be separated from the chiefly clayey mass. Furthermore, loess contains considerable amounts of calcium carbonate.

Even on the tiniest piece of loess a certain texture can be observed; the loess is traversed partly by extremely fine, partly by coarser tubes, branching out like reeds and frequently covered with white rims of calcium carbonate, but apart from this the loess is very porous and without the compact structure of clay. Loess is completely lacking stratification and is inclined to cleave off vertically.'

Loess may dovetail with glacial deposits (see p. 189) and with eolian sand (as in southern Poland: E. M. Dowgiallo, 1965). The stratigraphic relations of loess with other deposits confirm its glacial, as opposed to interglacial, origins. In some areas, for instance, a loess sheet may split into two members at the drift border, one leaf underlying the drift and the upper one passing over it. The lower member attests eolian action before the advance of ice and the glacial maximum, and often shows glaciotectonic disturbance by the advancing ice; the upper member blanketed the drift exposed later during retreat of ice. Fossil soils in all stages of development may be interbedded with the loess and appear in exposures as darker humic layers (see, for example, J. Jersak, 1969). These palaeosols are taken to mark more humid climatic intervals separating the dry loess substages, and are invaluable in providing radiocarbon dates to clarify the stratigraphic record. Thus the stratigraphy of the Wisconsin loesses of Illinois as far back as about 40,000 years is now controlled by more than twenty radiocarbon dates.

(b) *Theories of the origin of loess*   The earliest studies of loess in Europe suggested a lacustrine or alluvial origin. The originator of the eolian hypothesis was von Richthofen (1877), who based his arguments on studies of the China loesses; he was supported by R. Pumpelly (1879) who adopted the eolian hypothesis for the loess of central North America, and added the suggestion that the loess here was largely derived from glacial outwash. Controversy continued, however, and many continued to regard certain loesses as water-lain, especially those containing stratified silts. Recent opponents of the eolian hypothesis have included R. J. Russell (1944) and H. N. Fisk (1951), both investigating the loess of the lower Mississippi valley. Russell regarded the loess here as of fluvial-colluvial origin, derived from floodplain alluvial deposits by eluviation of the clays, and later attempted to apply the idea to Rhine valley loess. The majority opinion does not support Russell (see C. D. Holmes, 1944, and reply by Russell following this), and M. M. Leighton and H. B. Willman (1950) and others have marshalled an impressive array of evidence testifying to the eolian nature of Mississippi loess, even in the lower valley. However, the dispute has been of value, for it has stressed the possibility that certain formations labelled loess may not be eolian. H. Aumen and others (1965) reached this conclusion for some layers in the 'limon' deposits of Provence which show definite signs of water sorting. In the Soviet Union, non-eolian theories have also been propounded from time to time; for a recent exposition of the possible edaphic or eluvial origin of loess, see L. S. Berg (1960).

A perennial problem is that of distinguishing between primary (eolian) loess and secondary loess resulting from hill wash or slipping of the primary loess. Inevitably one may be confused with the other and mapping of primary loess distributions becomes very difficult. Thickness measurements will be unreliable—indeed, the immense thick-

nesses sometimes quoted for loess in China certainly include much material which is
not primary loess.

The classical theory explaining the origin of periglacial loess (see, for example, T. C.
Chamberlin, 1897), which is now generally and widely accepted, envisages the main
sources of wind-blown silt as glacial outwash plains. Across these plains flowed braided
and heavily silt-laden meltwater streams, which flooded in spring and summer, and
shrank in the later part of the year to expose broad expanses of unvegetated sediment.
The sediment was replenished each year and thus provided rich sources of fine silt.
As the sediments dried out each autumn, strong or gusty winds winnowed out the finer
silts and rock-flour, raising great dust clouds which are today a common phenomenon
on the outwash plains of Alaska, Iceland or Spitsbergen. T. L. Péwé (1955) describes
how loess is presently being created from the Tanana River floodplain in Alaska by
winds picking up rock flour from the outwash. Some of the dust may also have come
from till plains possessing little or no vegetation under a cold dry climate. The dust
was eventually deposited in regions away from the ice where vegetation of grass or
even forest could bind the wind-blown silt together.

The evidence generally supporting this theory is abundant and persuasive. The
eolian nature of loess is confirmed by the way the deposit mantles varying bedrock
formations indifferently, and by its independence of altitude (in Europe, for instance,
it sweeps from near sea level up to 500 m and over). Moreover, loess is unstratified
and in places merges into or is interbedded with coarser deposits of definitely eolian
character, including ventifacts. Its deposition in dry form is evident, for when subsequently wetted and loaded, it subsides or shrinks, showing that the original loess was
neither soaked by water nor deposited by it (V. A. Obruchev, 1945). Its thickness
diminishes away from the source regions which can often be identified by mineralogical
analyses.

Further valuable insight into conditions of deposition is given by the fauna. Mollusca
may occur in huge numbers—the Peoria loess (late Wisconsin) of the Great Plains
of the United States sometimes contains 180,000 shells per cubic metre. The shells
are often small and delicate, so fragile that emplacement by running water is inconceiv-
able. They lived on the surface, died, and were gradually entombed by the dust. The
mollusca are of comparatively few species and are terrestrial forms, the only aquatic
species being those associated with streams that dry up periodically. The gastropods
throw further light on the climatic conditions prevailing at the time the loess was accumulating. In the Peoria loess, A. N. Leonard and J. C. Frye (1954) found no species
which could resist long periods of drought or prolonged high temperatures; most
required a moist organic cover on the ground surface and were suited by cooler conditions than now prevail in the Great Plains in summer. Near the main rivers, forest
probably flourished and was instrumental in aiding accumulation of thick loess here
(30 m or more), while a few kilometres away, mollusca and the remains of prairie animals suggest grassland with occasional woodland patches. Altogether, Leonard and
Frye think conditions during the loess deposition were more favourable in this area to
plant and animal life than they are today, and the idea of a barren dusty landscape in the
loess accumulation areas here must be dismissed. Moreover, such favourable conditions must have lasted throughout the whole period of Peoria loess deposition, and

have existed over thousands of square kilometres, from north Texas to South Dakota, and east into Ohio, though there were certainly not uniform conditions over the whole of this vast area. As analyses of the loess point to a source in the outwash of valleys draining from glacier ice to the north or in the Rockies, and the mollusca were suited to a cool environment, loess deposition evidently took place in glacial rather than interglacial periods. This tallies with the stratigraphical relationships of the loess to the drift sheets: the Peoria loess, for instance, is interbedded with the Wood-fordian till (radiocarbon-dated 22,000–12,500 BP) in Illinois.

The picture that emerges may be summarized thus. At times of glaciation, outwash plains and to a lesser extent till plains pulverized by frost action and bare of vegetation were swept by strong, gusty, cold winds, often blowing off the ice. Dust clouds wafted across the barren tundra zone marginal to the ice, and the dust was deposited beyond, where vegetation cover existed to arrest its movement and helped to bind it together.

2.3   Distribution of loess

(a) *Continental Europe and USSR*   The greater part of the loess in this region lies beyond the maximum limit of glaciation. Loess mantles thousands of square kilometres of low-land in eastern Europe and European Russia, but is seldom thick enough to develop its own topography. The maximum thickness is about 80 m in Rumania. The vast outwash spreads of the Bug, Dnieper, Don and Volga appear to have been the main sources. Eastward, loess reaches into central Asia to the foothills of the Pamirs and Tien Shan, and north-eastward into Siberia.

An important loess belt in central Europe stretches from Cracow past Breslau and Leipzig towards Köln (where it was first described in the 1840s), and lies between the limits of the Weichsel and Elster glaciations, along the northern flank of the Mittel-gebirge; glacial outwash in the Urstromtäler provided the main source of silt. Impor-tant extensions of this belt into unglaciated country run up the Rhine Rift Valley, and westward into Belgium and northern France (Fig. 7.3).

The main source of silt for loess in these western areas was probably the floor of the North Sea or Channel, exposed during the low sea levels of glacial times. The low sea level also had the effect of extending the European land-mass, thus enhancing the continentality of the climate in central Europe where dry and severely cold conditions accompanied loess deposition. V. Lozek (1968) describes how the loess of central Europe formed under a climate experiencing mean annual temperatures between $0°$ and $-3°C$; very cold dry winters prevailed, with warm humid summers, and the vegetation was dominantly steppe grassland at lower altitudes.

The limon deposits of the region around Rouen and Le Havre have been shown to have a complex stratigraphy and characteristics not entirely attributable to wind deposition alone (J.-P. Lautridou, 1968). In this respect they are representative of many other areas of marginal loess deposits. Although basically eolian (the silt was blown in from the Seine estuary area exposed during low sea level), they have been modified by permafrost and by sheet wash from snow melt. The latter is responsible for a bedded structure, which is lacking in true loess.

Loess is rare in north-west Europe (except Iceland), for this area was largely ice-covered in the last glaciation, and its relatively maritime climate in the late- and post-glacial probably precluded much wind action. Some loess is known in Sweden, though; A. Falk (1955) has mapped loess up to 70 cm thick in the Mora area (latitude 61°), and the larger deposits at Brattforsheden in Värmland have been known for some time because of their interesting relationships with neighbouring sand-dune systems (see p. 192). Most Swedish loess is thought to date from about 9000 BP (late Yoldia or early Ancylus time).

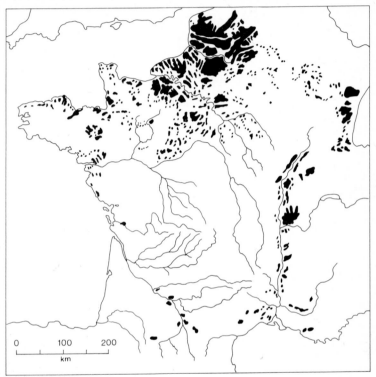

Fig. 7.3
Distribution of loess in France (F. Bordes, *Études françaises sur le Quaternaire*, 1969)

In summary, the principal features of loess in Europe are its concentration outside the limits of the Weichsel glaciations, its association with glacial outwash which provided the source areas, its more extensive occurrence in eastern rather than western Europe and on lowland rather than upland, and the clear indications of its formation under a dry continental climate during glacial periods.

The literature on the stratigraphy of loess and related formations in Europe and the Soviet Union is voluminous and no summary will be attempted here. Useful collections of papers reviewing the present state of knowledge for various regions are C. B. Schultz and J. C. Frye (1968) and *La stratigraphie des loess d'Europe* (1969).

(*b*) *Britain* Loess is rare in Britain apart from small patches in south-eastern England (for a full discussion of these see J. B. Dalrymple, 1960). The brick-earth of the London

Basin, southern East Anglia and Kent, comprises a silt loam of various origins, but much of it is probably a true wind-blown loess (S. W. Wooldridge and D. J. Smetham, 1931). The laminated variant seen in some sections probably represents deposition in small patches of standing water. Brickearth occurs up to elevations of 200 m on the Chilterns near Luton, where the inclusion of Palaeolithic implements shows its recent nature here. It is surprising that so few eolian deposits of substantial thickness have yet been found in and around the Thames valley, in which great masses of glacial outwash were laid down in the Saale and older glaciations. Perhaps the more maritime climate of Britain and the more rapid spread of vegetation were not so conducive to wind action as in continental Europe, and certainly hindered the drying-out of the surface of the outwash. The patches of loess in Britain are essentially of local origin, in contrast to the widespread loess of central and eastern Europe where powerful wind action operated over vast regions.

At Pegwell Bay, Kent, W. S. Pitcher and others (1954) have demonstrated the existence of a true eolian loess, derived from local Thanet Beds. Its age was later shown by M. P. Kerney (1965) to be the equivalent of the Older Coversand of Holland (Weichselian maximum); an overlying weak soil gave a radiocarbon date of about 13,000 years. In south-western England, also beyond the limits of glaciation, loess has only recently been recognized and more will certainly be revealed with further investigations of the superficial deposits. D. E. Coombe and L. C. Frost (1956) have examined the soils on the Lizard peninsula. These contain silt, probably derived in part from the Carnmenellis granite. In various samples, 43–57 per cent of the silt grains fall within the size range 0·01–0·05 mm. This in itself does not prove an eolian origin, but the wide scatter of silt in the region and its lack of addition today strongly suggest late Pleistocene wind action and the trapping of dust by vegetation. The silt overlies the early Pleistocene deposits of Crousa Common. On the Durham coast, C. T. Trechmann (1919) has claimed an interglacial age for loess 3–4 m thick banked against a north-facing slope. Its mineral composition was found to be remarkably similar to Rhine valley loess near Bonn. Only one eolian deposit is so far known from Scotland (R. W. Galloway, 1961), a silt 1 m thick near Kinross resting on and derived from local fluvio-glacial material.

(c) *North America*   There is an enormous literature on the loess deposits of North America, to which some references have already been made. A useful recent summary of Pleistocene eolian action in the United States is given by H. T. U. Smith (1964), and the distribution of loess is displayed in the map compiled by J. Thorp and others (1952). Stretching from the Rockies to Ohio, and from the lower Mississippi to beyond the Canadian border, most of the loess now seen is of Wisconsin (Weichselian) and later Illinoian age, for the older loess is often either denuded, buried, or decomposed too far to be easily recognizable. As in Europe, glacial outwash provided the chief sources, with the addition of some from bare till plains or non-glacial alluvium. Its distribution in relation to source areas reflects a dominantly westerly but variable air circulation, very similar to that of today, though probably stronger.

The best known occurrences of loess in the world lie in the Mississippi basin. In an area of 500,000 km², half of which is glaciated, are found most of the phenomena

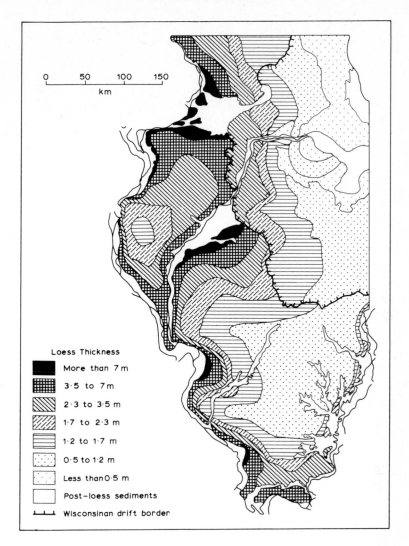

Fig. 7.4
Variations in the thickness of the Wisconsin loess in Illinois. Note the sudden changes at the Wisconsin drift border (R. F. Flint, *Glacial and Quaternary Geology*, John Wiley & Sons, New York, 1971)

associated with loess deposition (Leighton and Willman, 1950). Of particular interest is the Iowan loess (equivalent to the Morton loess of J. C. Frye *et al.*, 1962) and its unique relationship with the Iowan drift border. At the latter, the loess rises suddenly above the drift plain, looking rather like a moraine series, and attains a thickness of 15 m or more; it is also rather sandy, with some dune-like forms. Away from the drift border on the older Kansan drift, it gradually becomes finer in grain size and thinner. The loess is similar to the Iowan drift both chemically and mineralogically, ventifacts occur on the Iowan drift beneath a scanty loess cover, and beyond the drift border, the loess

mantles hills and valley slopes alike. Frye *et al.* (1962) have also successfully related
slight differences in mineral content of the various loess sheets to specific outwash-
carrying valleys from which the loesses were derived. Altogether, an eolian origin is
impossible to deny.

Figure 7.4 shows the variations in thickness of all Wisconsin loess in Illinois, and
demonstrates clearly the relationships of the loess to the Wisconsin ice limit and to
the major valleys carrying outwash. The loess is thickest along these valleys, especially
where they are widest; Leonard and Frye (1954) claim that a forested border along
the valley edges was important in trapping great thicknesses of loess here. Away from
the valleys, both thickness and average particle size of the loess diminish exponentially
with distance, again confirming the eolian origin (G. D. Smith, 1942). At the same
time, the carbonate content also decreases, in inverse proportion to the thickness of
the loess. Carbonate content at the time of deposition was over 40 per cent at the valley
bluffs where loess accumulated rapidly, but in the zones of thinner loess, deposition
was so slow that the loess was mostly leached of lime even when deposition had just
ceased.

A useful summary of the physical characteristics of Mississippi loess is given by E. L.
Krinitzsky and W. J. Turnbull (1967). Between 75 and 90 per cent of the loess falls
within the particle size range 5–75 μm; 10–25 per cent is less than 5 μm and less than
5 per cent exceeds 75 μm. Thicknesses are usually less than 3 m, but in one place attain
33 m. Close inspection has revealed faint signs of internal stratification, and it is suggested that periods of quiescence in wind action gave rise to some leaching of carbonates
and slight surface weathering. Mineralogical analyses show that calcareous types of
loess consist of 55–70 per cent quartz, but when carbonates are leached out, the proportion of quartz rises to 60–75 per cent and, in the oldest layers, to 80 per cent. The
properties of the Mississippi loess are essentially the same as those of other major world
loess accumulations: indeed, the similarities are remarkable. The principal
variations—clay and sand content, thickness and grain size—are wholly explicable in
terms of distance from source area and of secondary weathering.

Taken as a whole, the evidence is compelling that the great bulk of North American
loess is of eolian origin, and was laid down in glacial periods, with glacial outwash
providing the principal source of the material.

3 Niveo-eolian deposits

In the arid areas of present-day high polar regions, mixed accumulations of snow and
wind-blown sediment build up to thicknesses of a metre or more over periods of years.
A. Cailleux (1972) describes their occurrence in the McMurdo Sound region of Antarctica, and compares the deposits with those in a lower latitude (55°N) on Hudson Bay
where they are annual, the snow melting out each year. Wind ripples and miniature
dunes form on the surface, together with microforms of differential ablation such as
sand stalagmites beneath overhangs. In the Sudeten Mountains of south-west Poland,
A. Jahn (1972) records the progressive development of snow drifts and intercalated
dirt layers. Strong winter winds pick up both snow and sand or silt particles and deposit
them over the previous snow layer; then another snowfall covers this dirty layer and

a rhythmically stratified sequence is built up. As meltwater percolates down during any rise in temperature, it carries dirt with it, so that the dirt content increases in the lower older layers. In a snow drift at Mystów, three snow layers were separated by dark crusts of dirt:

Snow III	(February 1966):	dirt content	142 g/m³
Snow II	(January 1966):	dirt content	588 g/m³
Snow I	(December 1965):	dirt content	1600 g/m³

It is possible that, on the melting of a thick sequence of niveo-eolian deposits, a rhythmically-banded deposit similar to grèzes litées may result.

4 Sand dunes

In this context, we are concerned not with desert, coastal or lake-shore dunes of modern or Quaternary age, but with dunes now mostly inactive and stabilized that developed in Pleistocene periglacial regions. Such dune systems may yield valuable information on prevailing wind directions of the time, complementing the evidence of ventifacts or loess distribution; they sometimes point to conditions of greater aridity, and they may even reveal minor climatic fluctuations if dune building was intermittent. Interpretation of the evidence is, however, beset with difficulties and uncertainties, as H. T. U. Smith (1949) has emphasized.

Pleistocene dune systems of former periglacial regions are known from many parts of central North America and from Europe. I. Högbom (1923) was the first to present a unified picture of late Pleistocene dune building in Europe, although some of his interpretations have been modified by later work. Most dunes so far studied date from the last glaciation or the Late- and Post-glacial periods, for older ones have either been destroyed by subsequent ice advance, buried by loess, or weathered and eroded too far to be recognizable. They are closely related to glacial deposits, particularly outwash, which provided the main sources of sand, just as they provided the silt for loess. In north Germany, dunes are found along the Urstromtäler, and in Poland along the middle Vistula, for instance. The inland dunes of Poland have been intensively studied by A. Dylikowa (1969), R. Galon (1959) and others. Other sources of sand in the case of the European dunes included the sandy sea floors exposed by the lowered sea levels of glacial times, and in North America, lake floors exposed during the shrinking of such water bodies as Lake Agassiz and the glacial Great Lakes. The dunes do not necessarily indicate a dry climate during their period of activity—any conditions promoting surface desiccation will be favourable—but many workers have in fact concluded from all the available evidence that a drier climate was likely, or at any rate, could not be ruled out (for example, W. S. Cooper, 1935).

The forms of Pleistocene dunes vary greatly. The only major dune type not represented is the barchan, which modern studies of active desert dunes indicate requires complete absence of vegetation or other obstacles for most perfect development. In Pleistocene periglacial regions, vegetation appears to have been present to a limited degree, increasing its hold as the climate ameliorated but while the dunes were still active. Thus, complex dune forms occur. Their basic components are the longitudinal dune and the U-shaped ('parabolic' or 'transverse') dune (J. N. Jennings, 1957). The latter, with its horns pointing upwind, is not to be mistaken for the barchan. Examples of ancient parabolic dunes are given by F. Gullentops (1957). Determinations of former wind direction from dune morphology alone can be dangerous, especially if weathering has subdued the forms. Gross errors are possible if longitudinal and transverse dunes are confused (e.g. F. Enquist, 1932, who argued that certain Swedish dunes were longitudinal). In Europe, most Pleistocene dunes are of the parabolic variety. In the formation of parabolic dunes, sand from the inside faces is blown over the crest and deposited on the outer steeper leeward faces. The latter may slope at up to 35°.

Dune-sand bedding may be used to establish former wind directions with fair reliability if they are steeply dipping (30–35°) beds, for the true dip of these is always to the leeward in both U-shaped dunes and barchans. The method is less satisfactory if the dips are variable and less than 25°. One further point should be raised in connection with determinations of wind direction: the dune will record the direction of the effective sand-moving winds, which may or may not be that of the prevailing winds. Powerful storms from one quarter may shift far more sand in several hours than gentle prevailing winds from another quarter may achieve in several days or weeks.

An example of a Late-glacial dune-field in Sweden is described by F. Hjulström (1955). The dunes of Brattforsheden (Fig. 7.5) in Värmland cover about 10 km² and form a series of about twenty concentric curving ridges, convex towards the east and south-east, some of them subdividing and linking with others. They are probably transverse dunes and their forms suggest a north-westerly wind during their period of activity. F. Hjulström and A. Sundborg (1955) have suggested, however, that this does not represent the true direction of the isobaric gradient wind, for the surface winds may correspond to northerly or north-easterly gradient winds, and this is confirmed by the disposition of loess in the area, which has an eccentric position in relation to the dune field. This can be explained if the dunes were activated by a north-westerly surface wind but the loess was the result of deposition from a dust cloud blowing at higher levels in the atmosphere from the north or north-north-east. The hypothesis may help to account for some other anomalous loess distributions in central Europe, where the loess is often concentrated on east to north-east facing slopes, yet the dune systems indicate dominantly westerly to north-westerly winds.

The dune systems and loess of central Sweden demand strong and probably gusty winds from a generally northerly source; their age indicates the contemporaneous existence of ice to the north, and the open water of the Yoldia Sea or Ancylus Lake to the south. The strong temperature contrasts between ice and water favoured cyclogenesis, according to Sundborg, and squally northerly winds would prevail in the rear of each eastward-moving depression. The surface winds would descend from the ice cap to the lowlands of central Sweden, the resulting föhn effect helping to produce a low relative humidity to dry the surfaces of fluvioglacial deposits.

Dunes in northern Sweden and Finnish Lapland have been investigated by Seppälä (1971, 1972). In the former case, air photographs were used to identify dune systems, most of which consist of parabolic dunes. In northern Finland, 200 km² were studied in detail on the ground, and Fig. 7.6 exemplifies the intricate dune morphology. Grain-size analysis of the dune sands shows that they are particularly well sorted. Skewness

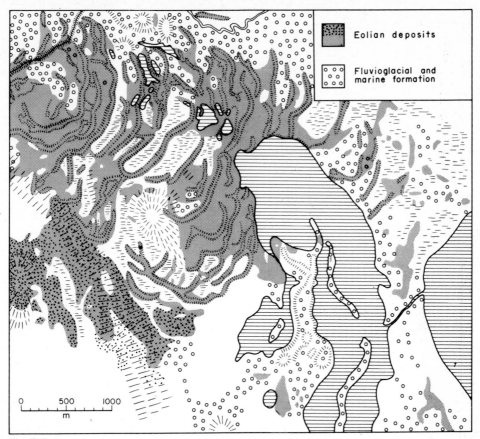

Fig. 7.5
Map of the eolian deposits in the northern part of Brattforsheden, Värmland, Sweden. The small dots indicate the loess (F. Hjulström, *Geogr. Annlr*, 1955)

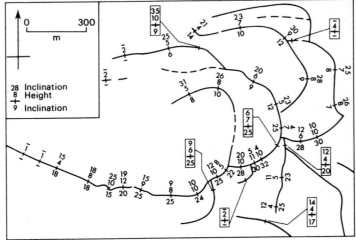

Fig. 7.6
Detailed morphology of a small area of Pleistocene sand dunes in Finnish Lapland (M. Seppälä, *Fennia*, 1971)

values are around 1·0 (showing that both the coarser and the finer fractions are equally well sorted), and kurtosis values are considerably greater than for glacial or most fluvioglacial deposits. The dunes enclose deflation basins which often become water-filled.

Because it is tempting to infer former wind directions from fossil dune systems and thereby reconstruct paleo-climates, there have been several attempts to make broad surveys of European Pleistocene dunes and to build up a picture of former wind and pressure systems. H. Poser (1948, 1950) argued for two main high pressure centres over Fennoscandia and central Europe (the Alps) respectively. The chief problem with such reconstructions is that the various dune systems are unlikely to be contemporaneous; and it has also been noted that there are considerable difficulties in inferring former wind directions from the dune forms themselves.

5 Conclusion

Periglacial wind action is associated with cold dry climatic conditions in regions where strong winds blow off nearby glacier ice, frost-shattered till plains and glacial outwash spreads. Wind erosion is manifested in the occurrence of grooved and polished bedrock surfaces and ventifacts. The latter if large and undisturbed can be used to establish former wind directions.

Disseminated eolian deposits consist of sand (0·06–1·00 mm) and silt (0·004–0·06 mm). Wind-blown sand grains are distinguishable by their roundness and possibly also by their frosted appearance, and occur in greatest abundance just beyond the last major Pleistocene ice limits. Silt-sized particles were blown in dust clouds farther from the ice and were redeposited as loess, the bulkiest of all periglacial accumulations. The eolian origin of primary loess is now firmly established, but within this broad category, it may be possible to differentiate between periglacial loess, with a narrow range of grain size since it was mostly derived from already-sorted fluvioglacial deposits, and desert loess of warmer climates. Periglacial loess is characterized by a dominance of grain sizes in the range 2–60 μm. It is unstratified, and its texture, faunal content, variations in thickness and relations with contemporaneous glacial deposits testify to its eolian origin and its association with nearby glacier ice. It mantles huge areas in eastern Europe, European Russia and North America, while smaller occurrences are being steadily revealed in Britain.

Fossilized dune systems formed under Pleistocene periglacial conditions are known in many parts of central North America and Europe. The commonest dune forms are parabolic (transverse) with horns pointing upwind, and longitudinal, and it is important to distinguish between the two types. The dune systems may then be used in some instances to reconstruct former wind directions, as in the case of some wind-eroded bedrock surfaces and large ventifacts, thus helping to add to our knowledge of Pleistocene meteorology.

6 References

La stratigraphie des loess d'Europe (1969), *Suppl. Bull. Ass. française Étude Quat.*, Paris, 176 pp.

AUMEN, H. *et al.* (1965), 'Petrographie des limons de Provence', *Bull. Ass. française Étude Quat.* **2**, 35–49

BAGNOLD, R. A. (1941), *The physics of blown sand and desert dunes*, 265 pp.

BERG, L. S. (1960), 'Loess as a product of weathering and soil formation', *Acad. Sci. U.S.S.R.*, Moscow (translated 1964, Israel program for scientific translations, Jerusalem), 216 pp.

BORDES, F. (1969), 'Le loess en France', *Études françaises sur le Quaternaire (Ass. française Étude Quat., Paris)*, 69–76

BRYAN, K. (1945), 'Glacial versus desert origin of loess', *Am. J. Sci.* **243**, 245–8

CAILLEUX, A. (1942), 'Les actions éoliennes périglaciaires en Europe', *Mém. Soc. géol. Fr.* **46**, 1–166

— (1969), 'Quaternary periglacial wind-worn sand grains in USSR' in *The periglacial environment: past and present* (ed. T. L. PÉWÉ), 285–301

— (1972), 'Les formes et dépôts nivéo-éoliens actuels en Antarctique et au Nouveau-Québec', *Cah. Géogr. Québ.* **16**, 377–409

CHAMBERLIN, T. C. (1897), 'Supplementary hypothesis respecting the origin of the loess of the Mississippi valley', *J. Geol.* **5**, 795–802

COOMBE, D. E. and FROST, L. C. (1956), 'The nature and origin of the soils over the Cornish serpentine', *J. Ecol.* **44**, 605–15

COOPER, W. S. (1935), 'The history of the upper Mississippi River in Late Wisconsin and Post-glacial time', *Bull. Minn. geol. Surv.* **26**, 72–108

DALRYMPLE, J. B. (1960), 'The nature, origin, age, and correlation of some of the brick-earths and associated soils of south-eastern England', Unpubl. Ph.D. thesis, Univ. of London

DOWGIALLO, E. M. (1965), 'Mutual relation between loess and dune accumulation in southern Poland', *Geogr. Polonica* **6**, 105–15

DYLIK, J. (1969), 'L'action du vent pendant le dernier âge froid sur le territoire de la Pologne centrale', *Biul. Peryglac.* **20**, 29–44

DYLIKOWA, A. (1969), 'Le problème des dunes intérieures en Pologne à la lumière des études de structure', *Biul. Peryglac.* **20**, 45–80

ENQVIST, F. (1932), 'The relation between dune-form and wind direction', *Geol. För. Stockh. Förh.* **54**, 19–59

FALK, A. (1955), 'Preliminary mapping of some localities of wind-blown silt in Dalarna', *Geogr. Annlr* **37**, 112–17

FISK, H. N. (1951), 'Loess and Quaternary geology of the lower Mississippi valley', *J. Geol.* **59**, 333–56

FRYE, J. C., GLASS, H. D. and WILLMAN, H. B. (1962), 'Stratigraphy and mineralogy of the Wisconsinan loesses of Illinois', *Circ. Ill. St. geol. Surv.* **334**

GALLOWAY, R. W. (1961), 'Periglacial phenomena in Scotland', *Geogr. Annlr* **43**, 348–53

GALON, R. (1959), 'New investigations of inland dunes in Poland', *Przeglad Geogr.* **31**, 93–110

GULLENTOPS, F. (1957), 'Quelques phénomènes géomorphologiques depuis le Pleni-Würm', *Bull. Soc. belge Géol. Paléont. Hydrol.* **66**, 86–95

HJULSTRÖM, F. (1955), 'The problem of the geographic location of wind-blown silt: an attempt of explanation', *Geogr. Annlr* **37**, 86–93

HÖGBOM, I. (1923), 'Ancient inland dunes of northern and middle Europe', *Geogr. Annlr* **5**, 113–242

HOLMES, C. D. (1944), 'Origin of loess—a criticism', *Am. J. Sci.* **242**, 442–6

JAHN, A. (1972), 'Niveo-eolian processes in the Sudetes Mountains', *Geogr. Polonica* **23**, 93–110

JENNINGS, J. N. (1957), 'On the orientation of parabolic or U-dunes', *Geogrl J.* **123**, 474–80

JERSAK, J. (1959), 'La stratigraphie des loess en Pologne, concernant plus particulièrement le dernier étage froid', *Biul. Peryglac.* **20**, 99–131

JOHNSSON, G. (1958), 'Periglacial wind and frost erosion at Klågerup, S.W. Scania', *Geogr. Annlr* **40**, 232–43. See also by the same author a further article in *Geogr. Annlr* **44**, 378–404

KERNEY, M. P. (1965), 'Weichselian deposits in the Isle of Thanet, east Kent', *Proc. Geol. Ass.* **76**, 269–74

KRINITZSKY, E. L. and TURNBULL, W. J. (1967), 'Loess deposits of Mississippi', *Geol. Soc. Am. Spec. Pap.* **94**, 64 pp.

KUENEN, P. H. (1960), 'Experimental abrasion 4: eolian action', *J. Geol.* **68**, 427–49

KUENEN, P. H. and PERDOK, W. G. (1962), 'Experimental abrasion 5: frosting and defrosting of quartz grains', *J. Geol.* **70**, 648–58

LAUTRIDOU, J.-P. (1968), 'Les loess de Saint-Romain et de Mesnil-Esnard (Pays de Caux)', *Bull. Centre Géomorph. Caen* **2**, 54 pp.

LEIGHTON, M. M. and WILLMAN, H. B. (1950), 'Loess formations of the Mississippi valley', *J. Geol.* **58**, 599–623

LEONARD, A. B. and FRYE, J. C. (1954), 'Ecological conditions accompanying loess deposition in the Great Plains region', *J. Geol.* **62**, 399–404

LOZEK, V. (1968), 'The loess environment in Central Europe' in SCHULTZ, C. B. and FRYE, J. C., *op. cit.* (1968), 67–80

MAARLEVELD, G. C. (1960), 'Wind directions and cover sands in the Netherlands', *Biul. Peryglac.* **8**, 49–58

NICHOLS, R. L. (1969), 'Geomorphology of Inglefield Land, north Greenland', *Meddr Grønland* **188** (1), 109 pp.

OBRUCHEV, V. A. (1945), 'Loess types and their origin', *Am. J. Sci.* **243**, 256–62

PAEPE, R. and PISSART, A. (1969), 'Periglacial structures in the late Pleistocene stratigraphy of Belgium', *Biul. Peryglac.* **20**, 321–36

PELISEK, J. (1963), 'Pleistozäne Dünensande in der tschechoslowakischen sozialistischen Republik', *Eiszeitalter Gegenw.* **14**, 216–23

PÉWÉ, T. L. (1955), 'Origin of the upland silt near Fairbanks, Alaska', *Bull. geol. Soc. Am.* **66**, 699–724

PITCHER, W. S. *et al.* (1954), 'The loess of Pegwell Bay, Kent, and its associated frost soils', *Geol. Mag.* **91**, 308–14

POPOV, V. V. (ed.) (1959), *Loess of northern China* (Moscow; translated 1964, Israel program for scientific translations, Jerusalem), 142 pp.

POSER, H. (1948), 'Äolische Ablagerungen und Klima des Spätglazials in Mittel- und Westeuropa', *Naturwissenschaften* **35**, 269–76 and 307–12

(1950), 'Zur Rekonstruktion der spätglazialen Luftdruckverhältnisse in Mittel-
    und Westeuropa auf Grund der vorzeitlichen Binnendünen', *Erdkunde* **4**, 81–8

POWERS, W. E. (1936), 'The evidences of wind abrasion', *J. Geol.* **44**, 214–19

PUMPELLY, R. (1879), 'The relation of secular rock disintegration to loess, glacial
    drift, and rock basins', *Am. J. Sci.* **17**, 133–44

RICHTHOFEN, F. VON (1877), *China* (Berlin), **1**, 758 pp.

RUSSELL, R. J. (1944), 'Lower Mississippi valley loess', *Bull. geol. Soc. Am.* **55**, 1–40

SCHÖNHAGE, W. (1969), 'Notes on the ventifacts in the Netherlands', *Biul. Peryglac.*
    20, 355–60

SCHULTZ, C. B. and FRYE, J. C. (eds.) (1968), 'Loess and related eolian deposits of the
    world', *Proc. 7th Congr. int. Quat. Ass.* (Boulder, Colorado, 1965), **12**, 369 pp.

SEPPÄLÄ, M. (1969), 'On the grain size and roundness of wind-blown sands in
    Finland as compared with some central European samples', *Bull. geol. Soc.
    Finl.* **41**, 165–81

(1970), 'Location, morphology and orientation of inland dunes in northern
    Sweden', *Geogr. Annlr* **54A**, 85–104

(1971), 'Evolution of eolian relief of the Kaamasjoki–Kiellajoki river basin in
    Finnish Lapland', *Fennia* **104**, 1–88

SHARP, R. P. (1949), 'Pleistocene ventifacts east of the Big Horn Mountains,
    Wyoming', *J. Geol.* **57**, 175–95

SMITH, G. D. (1942), 'Illinois loess: variations in its properties and distribution', *Bull.
    Ill. agric. Exp. Stn.* **490**

SMITH, H. T. U. (1949), 'Physical effects of Pleistocene climatic changes in
    non-glaciated areas', *Bull. geol. Soc. Am.* **60**, 1485–516

(1964), 'Periglacial eolian phenomena in the United States', *Rep. 6th Conf. int.
    Ass. quatern. Res. (Warsaw, 1961)*, Łodz (1964), **4**, 177–86

SUNDBORG, A. (1955), 'Meteorological and climatological conditions for the genesis
    of aeolian sediments', *Geogr. Annlr* **37**, 94–111

SWINEFORD, A. and FRYE, J. C. (1945), 'A mechanical analysis of wind-blown dust
    compared with analyses of loess', *Am. J. Sci.* **243**, 249–55

THORP, J. *et al.* (1952), *Pleistocene eolian deposits of the United States, Alaska, and parts of
    Canada* (map, scale 1/2,500,000, publ. *Am. geol. Soc.*)

TRECHMANN, C. T. (1919), 'On a deposit of interglacial loess and some transported
    pre-glacial freshwater clays on the Durham coast', *Q. J. geol. Soc. Lond.* **75**,
    173–203

TROLL, C. (1944), 'Strukturböden, Solifluktion, und Frostklimate der Erde', *Geol.
    Rdsch.* **34**, 545–694 (English translation, *Snow Ice Permafrost Res. Establ.*)

WENTWORTH, C. K. and DICKEY, R. I. (1935), 'Ventifact localities in the United
    States', *J. Geol.* **43**, 97–104

WOOLDRIDGE, S. W. and SMETHAM, D. J. (1931), 'The glacial drifts of Essex and
    Hertfordshire, and their bearing upon the agricultural and historical
    geography of the region', *Geogrl J.* **78**, 243–69

WRIGHT, H. E. (1946), 'Sand grains and periglacial climate: a discussion', *J. Geol.*
    54, 200–5

Index